Émilie Du Châtelet and the Foundations of Physical Science

The centerpiece of Émilie Du Châtelet's philosophy of science is her *Foundations of Physics*, first published in 1740. The *Foundations* contains epistemology, metaphysics, methodology, mechanics, and physics, including such pressing issues of the time as whether there are atoms, the appropriate roles of God and of hypotheses in scientific theorizing, how (if at all) bodies are capable of acting on one another, and whether gravity is an action-at-a-distance force. Du Châtelet sought to resolve these issues within a single philosophical framework that builds on her critique and appraisal of all the leading alternatives (Cartesian, Newtonian, Leibnizian, and so forth) of the period. The text is remarkable for being the first to attempt such a synthetic project, and even more so for the accessibility and clarity of the writing. This book argues that Du Châtelet put her finger on the central problems that lay at the intersection of physics and metaphysics at the time, and tackled them drawing on the most up-to-date resources available. It will be a useful source for students and scholars interested in the history and philosophy of science, and in the impact of women philosophers in the early modern period.

Katherine Brading is Professor of Philosophy at Duke University. She works primarily on philosophy of physics from the late 16th century to the present day. She is coeditor of *Symmetries in Physics: Philosophical Reflections* (2003).

Routledge Focus on Philosophy

Routledge Focus on Philosophy is an exciting and innovative new series, capturing and disseminating some of the best and most exciting new research in philosophy in short book form. Peer reviewed and at a maximum of fifty thousand words shorter than the typical research monograph, *Routledge Focus on Philosophy* titles are available in both ebook and print on demand format. Tackling big topics in a digestible format the series opens up important philosophical research for a wider audience, and as such is invaluable reading for the scholar, researcher and student seeking to keep their finger on the pulse of the discipline. The series also reflects the growing interdisciplinarity within philosophy and will be of interest to those in related disciplines across the humanities and social sciences.

Forthcoming titles:

Human Kinds
A Philosophical Defence
Marion Godman

Confucianism and the Philosophy of Well-Being
Richard Kim

The Philosophy and Psychology of Commitment
John Michael

Émilie Du Châtelet and the Foundations of Physical Science
Katherine Brading

https://www.routledge.com/Routledge-Focus-on-Philosophy/book-series/RFP

Émilie Du Châtelet and the Foundations of Physical Science

Katherine Brading

Routledge
Taylor & Francis Group
NEW YORK AND LONDON

First published 2019
by Routledge
52 Vanderbilt Avenue, New York, NY 10017

and by Routledge
2 Park Square, Milton Park, Abingdon, Oxon OX14 4RN

Routledge is an imprint of the Taylor & Francis Group, an informa business

Library of Congress Cataloging-in-Publication Data
Names: Brading, Katherine, 1970– author.
Title: Émilie du Châtelet and the foundations of physical science / Katherine Brading.
Description: New York: Routledge, 2019. | Series: Routledge focus on philosophy | Includes bibliographical references and index.
Identifiers: LCCN 2018051315 | ISBN 9781138351653 (hardback)
Subjects: LCSH: Du Châtelet, Gabrielle Emilie Le Tonnelier de Breteuil, marquise, 1706–1749. | Women physicists—France—Biography. | Physicists—France—Biography. | Women mathematicians—France—Biography. | Mathematicians—France—Biography. | Physics—Early works to 1800. | Mathematics—Early works to 1800. | Du Châtelet, Gabrielle Emilie Le Tonnelier de Breteuil, marquise, 1706–1749. Institutions de physique.
Classification: LCC QC16.D86 B73 2019 | DDC 530.092 [B]—dc23
LC record available at https://lccn.loc.gov/2018051315

ISBN: 978-1-138-35165-3 (hbk)
ISBN: 978-0-429-43517-1 (ebk)

Typeset in Times New Roman
by codeMantra

To my family
Mark, Thomas, and Matthew
Peter, Penelope, and Liz

Contents

Preface

In the spring of 2014, as I pondered the philosophical problems left unsolved in the wake of Newton's *Principia*, I picked up the *Institutions de Physique* of Gabrielle Émilie le Tonnelier de Breteuil, marquise Du Châtelet. There, I found a text that addressed exactly the questions I was interested in. Within a few weeks, a group of interested students and I had formed, including graduate students in the History and Philosophy of Science Program and the Philosophy Department, as well as a graduate student in French and an undergraduate student in French and History, at the University of Notre Dame. We set ourselves up as an official graduate seminar, with the goal of translating and reading the *Institutions* chapter by chapter, week by week. It was an exciting semester of adventure and discovery. By the end of it, we had a rough translation of almost all the chapters and sections not already translated by Isabel Bour and Judith Zinsser, and had read the entirety of the text cover to cover. We knew enough to know that this is an important text in the history of philosophy.

The seminar participants were Bohang Chen, Jamee Elder, John Hanson, Lynn Joy, Lauren (LaMore) Montes, Anne Seul, Phillip R. Sloan, Monica Solomon, Jeremy Steeger, Eric Watkins, and Aaron Wells, and we had visits from Andrew Janiak, Michela Massimi, Lydia Patton, and Marius Stan. We were excited to learn about Janiak's joint project on Du Châtelet with Karen Detlefsen and his plans for the as-yet unnamed Project Vox website. Something was in the air, and the moment seemed right. In April 2015, we followed up on the work of the previous semester with a workshop, to which we invited Judith Zinsser. I am grateful to Judith for her unstinting support from that event onwards. Fittingly, we concluded the April workshop by attending a Notre Dame production of Voltaire's *Candide*. By this time, evidence of wider interest in recovering Du Châtelet as a philosopher in her own right was growing, with workshops and conferences beginning to appear. When a moment such as this arises, we should seize it, and we should remember that such moments cannot arise without all the work that has gone before. Particularly important for Du Châtelet's philosophy in her *Institutions* is the pioneering work of Carolyn Iltis (1977) and Linda Gardiner Janik (1982) as well as more recent work by, among others, Sarah Hutton; Anne-Lise Rey; Karen Detlefsen; and Ruth Hagengruber, along with those who contributed to her edited volume (Hagengruber 2012). In addition, the first venues to welcome and promote scholarship on Du Châtelet were those which explicitly sought to recover lost female voices in the history of philosophy. I am grateful to all those who labored against the current, without whom there would have been neither the foundation on which to build today's Du Châtelet scholarship nor the venues at which to present it.[1]

This book grew out of work begun in the Notre Dame seminar described earlier and has been brought to completion while working with Marius Stan on our monograph on 18th-century matter theory and mechanics, during research leave funded by the American Council of Learned Societies. My thanks to those who provided feedback on aspects of this project and chapters of the manuscript, especially Zvi Biener, Mary Domski, Bryce Gessell, Pablo Ruiz de Olano, Don Rutherford, Andrea Sangiacomo, Monica Solomon, Marius Stan, Katie Tabb, and the anonymous reviewers for Routledge. I am grateful to Andrew Weckenmann at Routledge for his encouragement and support, to Penelope Brading for her assistance in preparing the manuscript, to Jeanine Furino and her production team, and to Cheryl Thomas and Qiu Lin for their help with sources and references. Moreover, with work on Du Châtelet's philosophy progressing at such a pace, I have benefited from unpublished manuscripts by several people, including Karen Detlefsen, Bryce Gessell, John Hanson, Anne Seul, Monica Solomon, and Aaron Wells.

Research on Du Châtelet as a philosopher is at an early stage, and the work of integrating her into the history of philosophy as it is taught in our classrooms is only just beginning. In Anglo-American philosophy, such work is greatly facilitated by the existence of an English translation of the relevant texts. I am grateful to Isabel Bour and Judith Zinsser for their translation of multiple chapters and to Lydia Patton for her translation of Chapter 9. For the remaining chapters, I am grateful to the seminar participants, and especially to my mother, Penelope Brading, for all the work that has gone into making a complete English translation of the 1740 first edition of the *Institutions de Physique* available.

Note: To differentiate the references to chapters within this text from those in Du Châtelet's, we have noted chapter references to this book in italic, while references to chapters within Du Châtelet's book are roman.

Note

1 Personal thanks to Eileen O'Neill's "Disappearing Ink: Early Modern Women Philosophers and Their Fate in History" (1998), which makes vivid the manifold ways in which the contributions of so many early modern women philosophers were lost. It is now becoming more widely accepted that the Western philosophical canon is a product of our history in ways that demand a more sophisticated treatment in our retelling of our philosophical inheritance.

1 Introduction

The centerpiece of Émilie Du Châtelet's philosophy is her *Institutions de Physique* (hereafter her *Foundations of Physics* or simply her *Foundations*), first published in 1740. My purpose here is to offer an introduction to the philosophy found within.[1] The *Foundations* contains epistemology, metaphysics, methodology, mechanics, and physics, including such pressing issues of the time as whether matter is infinitely divisible, the appropriate roles of God and of hypotheses in scientific theorizing, and whether gravity is an action-at-a-distance force. Du Châtelet sought to resolve these issues within a single philosophical framework, building on her critique and appraisal of all the leading alternatives of the period (Cartesian, Newtonian, Leibnizian, and so forth). The text is remarkable for being the first to attempt such a synthetic project. Du Châtelet put her finger on the central problems that lay at the intersection of physics and metaphysics at the time, and tackled them using the most up-to-date resources available.

With such a wide range of philosophical topics being covered in a relatively short text, it would be easy to pick on particular topics and criticize Du Châtelet's treatment of them individually. Such criticisms are surely important for the long-term evaluation of the text, but first things first. The *Foundations* has yet to receive detailed treatment as a work of philosophy, and we have yet to establish the overall philosophical goals of the book. Our first reading of the arguments should be with respect to these goals. And so, our first question must be: what was Du Châtelet trying to achieve?

Du Châtelet tells us in her Preface that there was, at the time she was writing, no complete and up-to-date textbook in physics available in French, and that her aim is to rectify this by making material from Newton and Leibniz available in French. By itself, this would give a descriptive goal to the project, reporting the work of others, but Du Châtelet makes choices about what materials to include (and exclude), and she puts these materials to work in a philosophical project of her own. According to the reading of the *Foundations* that I offer in this book, the central physical and metaphysical problem of the text concerns how it is that bodies act upon one another, and the central epistemological problem concerns the appropriate method by which to address this problem of bodily action. These are the goals with respect to which we should read and assess the text.

In what follows, I outline my view of Du Châtelet's goals in the *Foundations of Physics* and my preferred reading of the text, to be developed in more detail in *Chapters 2–4*. In Section 1.1, I provide some background to the *Foundations* and some guidance on how to read it. This section draws on the existing secondary literature, which I discuss in more detail in Section 1.2. In Section 1.3, I provide an extended overview of the remainder of this book, which can be summarized as follows.

In *Chapter 2*, I show that the problem of method plays a crucial role in the early chapters of the *Foundations*, and I describe Du Châtelet's innovative two-pronged methodology for scientific theorizing. Her introduction of this methodology was motivated by her interest in bodily action, including Newtonian action-at-a-distance, *vis viva*, and human agency. In *Chapter 3*, I discuss Du Châtelet's detailed account of matter, bodies, and force, and in *Chapter 4*, I present her attempts to address several controversial issues of bodily action using the resources set out in *Chapters 2* and *3*. Our understanding of her approach to these issues depends upon our understanding of her method and of her account of matter, body, and force; and reciprocally our understanding of these resources, as she develops them in her text, is deepened by our understanding of the problems they are intended to help solve. By framing our philosophical engagement with Du Châtelet's *Foundations of Physics* in the way that I propose, we are able to do justice to the philosophical richness of the text and to its significance for history of philosophy.

1.1 Reading the *Foundations*: Some Background and Context

The text that I discuss in this book is the first edition of the *Foundations of Physics*, published in 1740.[2] A second edition appeared in 1742, which was translated into both German and Italian (in 1743).[3] Crucial passages were reproduced (often without attribution) in the *Encyclopédie* (hereafter *Encyclopedia*) of Diderot and d'Alembert (publication of which began in 1751).[4] A partial manuscript version of the text is extant, providing evidence of substantial changes shortly before publication (about which more later). For a deeper understanding of the philosophy of the *Foundations,* much work remains to be done to study the philosophical significance of changes between the manuscript, the 1740 edition, and the 1742 second edition, as well as comparisons with the German and Italian translations and among variants of the editions. In this book, I am writing only about the 1740 published edition (with occasional references to the manuscript).

The existing English-language secondary literature on the *Foundations* is small (for more details, see Section 1.2 and references throughout). Three areas require more detailed research. First, work is needed on the differences between the multiple versions of the text, as noted. Second, there is a great deal more work to be done in studying the philosophical context, sources, and influence of the *Foundations.* Third, the study of Du Châtelet's corpus as a whole, investigating the philosophical relationships among the texts, is a project as yet barely begun. My discussion of the *Foundations of Physics* is inevitably impoverished due to the early stage of development of these three areas of research. My hope is that the account of the *Foundations* offered here, despite its limitations, will prove useful for further advancing the scholarship on Du Châtelet's philosophy.

In this section, I offer some background and context that help us understand how to read the *Foundations.* In the history of philosophy, Du Châtelet's *Foundations of Physics* is situated after Locke's *Essay* (1689); Newton's *Principia* (in the three editions of 1687, 1713, 1726; see Newton, 1999); *The Leibniz-Clarke Correspondence* (published in 1717; see Leibniz and Clarke, 1998); and Berkeley's *Principles* (1710), *Dialogues* (1713), and *De Motu* (1721). It is contemporaneous with Hume's *Treatise* (1739–40), and it comes before Hume's *An Enquiry Concerning Human Understanding* (first published under a slightly different title in 1748) as well as before Kant's earliest publication, in 1749, on living forces (Kant, 2012b), in which he references Du Châtelet. Philosophically, Du Châtelet's primary concern is with bodily causation: with how it is that bodies are capable of acting causally upon one another, and with how we can know that they do. As we will see, this concern takes her into a broad range of issues familiar from the philosophy of the period.

It is widely acknowledged that Du Châtelet assisted Voltaire in the writing of his *Éléments de la philosophie de Newton*, first published in 1738.[5] Du Châtelet published a review of this book (1738), and in the Preface to the *Foundations* she makes explicit her intention of going beyond the "narrow boundaries" of Voltaire's exposition (2009, 1.VI). Du Châtelet's work with him (including their experimental work at Cirey, her readings in philosophy and physics, her lessons in mathematics, her discussions with such figures as Algarotti, and her conversations and correspondence with Maupertuis) provides the intellectual context for her writing of the *Foundations*.[6] Her studies in philosophy included Plato's *Dialogues*, Locke's *Essay*, and *The Leibniz-Clarke Correspondence*, among other things.[7] Du Châtelet's engagement with Locke is rarely explicit in the *Foundations*, unlike her engagement with Descartes, Leibniz, Newton, and Wolff, but it is nevertheless a crucial presence.[8]

The first version of the *Foundations* was approved for publication on 18 September 1738. The "approbation" makes clear that the original work contained an exposition of "the principles of philosophy" of Leibniz and of Newton. It states (my translation)[9]:

> **Approbation**
> I have read, by order of the Lord Chancellor, a Manuscript which has for its title: *Foundations of Physics*, this Work in which are explained the principles of the Philosophy of Mr. Leibniz and those of Mr. Newton, is written with great clarity, and I have found in it nothing that could prevent printing. Paris, 18 September 1738. Signed Pitot.

Du Châtelet's review of Voltaire's *Elements* had appeared in that same month, to be followed soon after by her *Dissertation on the Nature and Propagation of Fire* (1739). Other work completed or in progress at the time includes her essays on optics and on liberty, and while not published during her lifetime both circulated in manuscript form.

An abrupt suspension of printing of the *Foundations* took place in the winter of 1738–39, and there has been much speculation as to the causes, not least since we do not have a complete version of the original manuscript.[10] Relevant factors include her ongoing discussions with Maupertuis concerning *vis viva* and her desire to learn more mathematics before proceeding.[11] Samuel König arrived in Cirey as a mathematics tutor in the spring of 1739,[12] and once Du Châtelet discovered that he had studied under Wolff, she enlisted his help "to summarize key parts of the chapters that she needed for her revisions" (Zinsser, 2006, p. 171). Whatever the motivation and original plans for revising the text, the upshot involved major changes to the opening chapters, drawing on the metaphysics of Leibniz and of Wolff.

Famously, König accused Du Châtelet of having copied from him in preparing the revised chapters of the *Foundations*. In my opinion, the evidence (from Du Châtelet's own prior reading, from her correspondence (especially in the late 1730s with Maupertuis), from the development of her philosophical positions through the period in question, and from König's own behavior and relationships) refutes any charge of plagiarism. I will not pursue this here.[13] Insofar as my work makes a contribution to this issue, it does so by demonstrating the philosophical agenda that runs through the entire text of the *Foundations*; by showing that this agenda predated the revisions; and by thereby making vivid the philosophical contributions of the author through the choices of the materials, the formulation that she gives, the changes that she makes, and the uses to which she puts the materials in the arguments that she makes in the text.

The *Foundations* was resubmitted for publication in September 1740 and appeared in print in December of that year. The immediate reception of the text is discussed by Janik (1982, pp. 97–8) and by Zinsser (2006, p. 191). Du Châtelet had intended anonymous publication, in order to get an impartial response to the book, but this intention was thwarted by König who revealed her authorship shortly before publication. Nevertheless, Du Châtelet sent copies of the book to leading figures, and it was positively reviewed in the *Journal des sçavans* (in two parts, the first in December 1740 and the second in March 1741), with extensive quotations. The ongoing influence of the text, both direct and indirect (e.g. through extracts reproduced in the *Encyclopedia* of Diderot and d'Alembert; see earlier), requires further research. Suisky (2012, p. 153) offers the following assessment: "The extraordinary role Du Châtelet's *Institutions de physique* played in the eighteenth century is... confirmed by... the esteem her treatise met in the public that was probably surpassed only by Euler's *Lettres à une princesse d'Allemagne*".

In late April 1740, Du Châtelet wrote in a letter to Frederick of Prussia that the idea of writing such a book began with lessons that she prepared for her young son, Florent-Louis, who was ten years old at the time (see Zinsser, 2006, p. 165). Of this, Wade (1969, p. 277) writes, "Although Mme du Châtelet adopted early the fiction of an essay on physics written for her son, in reality what she was doing was producing a text on physics". Who, then, was her audience, and how should we situate her text within the literature of the time?[14]

I think it is clear that Du Châtelet did not see herself as writing in the genre of Fontanelle and Algarotti, who wrote "for the ladies" and whose work she viewed as lightweight. I agree with Harth's assessment (1992, p. 202; see also Hutton, 2012, p. 85):

> It is more likely that she thought of herself as a successor to Rohault in the Cartesian enterprise of writing serious but

> comprehensible philosophy. Her stated goal was not to convert her public to the party of philosophy, as Fontanelle wished to do, but rather to add to the stock of knowledge in a collective search for truth.

That being said, I also agree with Kawashima (2004) that Du Châtelet did not model her text on either Rohault's textbook or Descartes's *Principles of Philosophy* (though see Detlefsen, 2014 and forthcoming). Rather, the structure and list of topics is that found in the Newtonian texts explicitly recommended by Voltaire in his *Elements*: books by Keill, Musschenbroek, 's Gravesande, and Pemberton (see Appendix 1 for details). I take these to have been her model.[15] As we shall see in *Chapter 2*, all of these texts begin with considerations of method. The topics covered thereafter match those of the *Foundations* (see Appendix 1).[16] I believe that it is appropriate to approach Du Châtelet's *Foundations* as being primarily intended to provide French readership with a book of physics modeled on these Latin- and English-language texts. The most striking difference between these Newtonian texts and the *Foundations* is Du Châtelet's inclusion of metaphysical commitments, for which she explicitly refers to Leibniz and Wolff.[17]

It is all too easy, however, to read the terms "physics" and "metaphysics" with their 21st-century meanings and thereby fall into misunderstandings.[18] Du Châtelet uses the terminology of physics and metaphysics to mark a distinction, but we must not assume that where we might tend to draw the line between them is where Du Châtelet saw the line fall. Hazards abound, exacerbated by reading a French text in English translation. The term "physics" did not mean then what it means now. An alternative English phrase in use at the time might be "natural philosophy" as a translation of "physique", but since there was disagreement at the time over what falls under the remit of natural philosophy, this label does not resolve our problems. In the face of this difficulty, I think that Musschenbroek's characterization is helpful. According to Musschenbroek (1744),[19]

> [Physics] considers the space of the whole universe, and all bodies contained in it; enquires into their nature, attributes, properties, actions, passions, situation, order, powers, causes, effects, modes, magnitudes, origins; proving these mathematically, as far as may be done

whereas

> [Metaphysics] explains such general things as are common to all created beings. As what is being, substance, mode, relation, possible, impossible, necessary, contingent, etc.

On this division of responsibilities, Du Châtelet's text is primarily about physics: it is about the nature of bodies and their properties, actions, and passions; and, most importantly, it is about *causes*. Indeed, physics (as understood at the time) was most importantly a search for *causes*, and a successful physics provided causal knowledge concerning the behaviors of bodies. In titling her book *Foundations of Physics*, Du Châtelet signaled her concern with causal knowledge.[20]

With this information in hand, we should expect the method that she develops in her early chapters to be directed toward securing knowledge of causes, and this is exactly what we find, as we shall see. We should also expect Du Châtelet to draw in metaphysics only where it is needed in order to address questions arising in physics. I believe that this is how we should approach the metaphysics that we find in the text, including the metaphysics that Du Châtelet introduced in her revisions to the manuscript.

In the reading of the *Foundations* offered in this book, the problem of bodily action takes center stage. Approached in this way, the structure of the *Foundations* is as follows (see Appendix 1 for a list of chapter headings). Following the Preface, Du Châtelet offers four chapters that provide a framework for knowledge of bodies and bodily action. The topics covered in Chapters 1–4 are the principles of our knowledge; the existence of God; essences, attributes, and modes; and hypotheses. These chapters should be read as primarily methodological in intent: they concern the methods by which we can secure such knowledge. Chapters 5–10 deal with controversial issues concerning space, time, and matter, using resources from Chapters 1–4. Chapters 5 and 6 are on space and time, respectively, as prerequisites for an account of extended bodies and their motions, while Chapters 7–10 provide Du Châtelet's account of the bodies that are the subject matter of physics such that these bodies are capable of acting on one another. Chapters 11–21 discuss the motions of these bodies. Chapters 11 and 12 include the laws of motion, while Chapters 13–19 deal with gravitation, including results on the motion of bodies under the influence of gravity developed by Galileo, Huygens, and Newton, among others. Newton's theory of universal gravitation was both new (and so not covered by Rohault) and controversial (in its claims about action-at-a-distance), and the problem of bodily action posed by Newtonian attraction is treated in Chapter 16. The final two chapters of the *Foundations* concern controversial issues surrounding contact action (Chapter 20) and the force of bodies in motion (Chapter 21), the latter being Du Châtelet's contribution to the so-called "*vis viva* controversy". Seen in this way, the *Foundations* follows the structure of prior Newtonian textbooks (see Appendix 1), notwithstanding its sometimes unusual content in comparison to these earlier texts.

The key to reading Du Châtelet's book as a work of philosophy, I will argue, is to approach it in three steps, beginning with the problem of method for natural philosophy, and moving from there to her account of matter, body, and force, and finally to her application of these resources in tackling key problems in the physics of bodies of the time. Approached in this way, the text reads as a sustained argument concerning bodily action.

In my view, the *Foundations* thereby comprises a unified whole. The modern reader might, however, be tempted to see a disunified amalgam of metaphysics and physics. Such a reader would find their view supported by some of the claims in the secondary literature. In the next section I review the secondary literature and continue my argument for my preferred approach to the text.

1.2 Reading the *Foundations*: The "Received View" and Beyond

The "Received View" of Du Châtelet's *Foundations* is that it presents a "marriage between Leibnizian metaphysics and Newtonian science", with Leibnizian metaphysics providing "the metaphysical foundation which in her view was an essential pre-requisite for scientific thinking" (Barber, 2006 [1967], p. 22).[21] This approach is reflected in the recent scholarship that has done so much toward recovering Du Châtelet as a philosopher, including Iltis (1977), Janik (1982), and some contributions to the Hagengruber (2012a) edited collection.[22]

Nevertheless, I think the approach is problematic. Such a view encourages us to approach the text with the question "Does Du Châtelet succeed in providing a Leibnizian foundation for Newtonian physics?", and I think this is the wrong question to be asking as our first question when we approach the text, because this was not Du Châtelet's primary objective. So, while this may be an important question to ask about the text, we should tread carefully: with work on Du Châtelet by philosophers at such an early stage, we should begin by treating the text on its own terms. This is what I try to do in this book.[23]

One danger of the "Received View" is that it invites an interpretation of Du Châtelet as passive with respect to the philosophical content of her book, copying ideas from Leibniz, Wolff, and Newton, and placing them within the covers of a single book. An example of this tendency is seen in Barber's (2006 [1967]) paper on the genesis of the *Foundations* when he writes (p. 23):

> Remarkable in its historical context as her achievement is, we should perhaps accept her own verdict that it was by means of translation and exposition rather than original work that she was best equipped to help the cause of enlightenment. It seems unlikely, then, that in this sphere she can have exerted any real influence on Voltaire's thought.

Were this correct, then the appropriate way to evaluate the text would be on its merits as a translation and exposition. There is some textual evidence for the view that in the *Foundations* Du Châtelet was translating the views of Leibniz and Newton into French, in order to make them available to a French audience for the first time. As noted, Du Châtelet tells us in her Preface that there is no complete and up-to-date book of physics in French. The Cartesian textbook by Rohault was, she says, very good for its time, but is now inadequate due to the many discoveries made since its publication decades earlier.[24] The work of Descartes being readily available in France at that time, her aim is to make material from Newton and Leibniz available in French, too. Du Châtelet says that she plans to gather in one place materials already available elsewhere in Latin, Italian, and English (Du Châtelet, 2009, Preface, III); that she will make her reader acquainted with the system of Newton (Preface, VI); and that she will explain the principal opinions of Leibniz, drawing them from Wolff (Preface, XII). By itself, this would give a descriptive goal to the project: reporting the work of others. If we approach the text in this way, then it is natural to view it as an amalgam of two halves, one half an exposition of Leibnizian metaphysics and the other half of Newtonian physics, with little if any connection between them. It would also be natural to evaluate the text according to how faithfully Du Châtelet reported the opinions of her sources. Though natural, these would be mistakes, as more recent work has made clear.

Wade began the movement toward a different approach when he wrote (1969, p. 279):

> Her book, which was designed to trace the activities of modern physics, did everything possible to incorporate the ideas of Descartes, Leibniz, and Newton and to describe the role each played in the establishment of modern physics.

This approach draws our attention to Du Châtelet's active engagement with her source texts: she had to decide which aspects of the philosophy of each figure to incorporate and integrate as contributions, and which to discard. Following Wade's lead, recent secondary literature evaluates the *Foundations* as an active synthesis of materials drawn from Newton, Leibniz, and Wolff, among others, with Iltis (1977) and Janik (1982) publishing groundbreaking papers on the content of the *Foundations*. Iltis stresses the innovative character of Du Châtelet's project as an attempt to integrate resources from Descartes, Leibniz, and Newton, writing (1977, p. 31):

> Although Newtonian in its basic mechanical principles, the resulting work followed Leibniz on the subject of dynamics, while the natural philosophy of the early chapters presented an integration of elements from the thought of Leibniz, Descartes, and Newton.
>
> It is this integrative character of Madame du Châtelet's thought which sets her *Institutions* apart from other attempts to disseminate Newtonian mechanics.

Similarly, Janik (1982, p. 98) notes that in the 1740s, Du Châtelet's *Foundations* was the *only* text that argued "for the compatibility of the basic Wolffian doctrines with Newtonian physics". Janik reports that this feature of Du Châtelet's text was largely missed, suggesting that many read only the early chapters along with the final chapter on *vis viva*, and failed to realize that the intervening chapters presented "a reinterpretation of Newton on the basis of a thoroughly Wolffian ontology, relying heavily on the principle of sufficient reason to justify both Newton's empirical discoveries and their metaphysical necessity". I agree with Janik's assessment that Du Châtelet was the first to recognize that Leibnizian and Wolffian resources could be used to address the metaphysical questions that are raised by Newtonian physics but which the Newtonians themselves lacked the resources to answer, and I support her view, now widely accepted, that the originality of Du Châtelet's synthesizing project cannot be doubted. Though I disagree with other aspects of Janik's treatment and offer an alternative approach to the text, I also maintain that her paper is essential reading for anyone interested in the *Foundations* as a philosophical text. It is, at the time of writing of this book, the single most important source on the philosophy of the *Foundations*.[25]

The synthetic character of the *Foundations* is, I think, beyond dispute. It is clear, therefore, that the text goes beyond translation and exposition. Du Châtelet identified what she took to be the deepest problems facing natural philosophy at the time and attempted to mobilize the resources she found – in the Cartesians, the Newtonians, the Leibnizians, and so forth – to address these problems. Any reader of the *Foundations*, whether primarily Cartesian, Leibnizian, or Newtonian in inclination, would have recognized elements of their own position that are endorsed by Du Châtelet, and elements that she shows face serious difficulties. The *Foundations* is a highly perceptive review of the state of natural philosophy in 1740 and the key problems therein, and this in itself makes it an important text in the history of philosophy.

However, in my assessment, the *Foundations* offers more than its characterization as a synthesis might imply. In the process of addressing the problems that she took to be most pressing, Du Châtelet transformed the resources on which she drew, and in so doing, she developed new positions and transformed the philosophical landscape surrounding the theory of matter, as it stood in 1740. This positive assessment of the philosophical content of the *Foundations* stands at odds with the assessments of Iltis (1977) and Janik (1982), and our disagreement has its source, I believe, in the way in which the "Received View" frames the text.

In line with the "Received View", Iltis frames the text as an attempt to provide Leibnizian metaphysical foundations (in the early chapters) for Newtonian physics (as presented in the later chapters) and evaluates the philosophical content of the early chapters within this framework. Despite her praise for the integrative goals of the project, in the end she arrives at a negative assessment of the philosophical merits of the text.[26] Iltis's negative evaluation seems to arise from looking at Du Châtelet's text to see whether she got Leibniz right, rather than beginning with the problems Du Châtelet herself was trying to solve, and then looking to see how far she got using whatever resources she happened to draw on (be they "Leibnizian" or otherwise). If we consider the early chapters in isolation from the overall project of the *Foundations*, as the "Received View" encourages us to do, we fail to assess them with respect to their philosophical purposes. It is my contention that closer attention to Du Châtelet's own goals in the *Foundations*, and to the details of her arguments, yields a more positive assessment.

Janik (1982, p. 93) presents what I have called the "Received View" thus:

> Studying the *Institutions* makes it clear that one of her basic convictions, from the beginning of its composition, was that science, whether Newtonian or otherwise, was dangerously incomplete without some kind of metaphysical foundation.

I think this is not quite right. Du Châtelet did come to believe that Newtonian physics is dangerously incomplete. She also came to believe that it could be completed by developing resources found in Leibniz and Wolff. It is also true that in drawing on these resources, she introduced significant metaphysical commitments not found in earlier textbooks on Newtonian physics. Nevertheless, the incompleteness of Newtonian physics is not metaphysical in the first instance. The basic foundational problem addressed by Du Châtelet is the lack of an epistemically secure basis for physics, and her response is to propose a new method for arriving at scientific knowledge. Thus, when she writes to Maupertuis (23 October 1734, my translation), "I divide my time between the builders and Mr. Locke, for I seek the foundation of things for one just as for the other", we must not assume that by this she means a metaphysical foundation for Locke's philosophy. It may be that a secure basis requires the introduction of metaphysics, but the motivation is not primarily that of providing a "metaphysical foundation" for physics.[27]

I think that regarding Du Châtelet's search for foundations not as *primarily* the search for metaphysical foundations, but rather as a quest for a secure basis for natural philosophy more generally, has implications for Janik's view of Du Châtelet's overall project. Janik (1982, p. 109) writes:

> In the final analysis, for mme Du Châtelet the justification of scientific knowledge via the principle of sufficient reason was a project to be undertaken with extreme caution and self-criticism, constantly subjected to correction, likely never to reach more than a tentative and elementary stage, and in practice irrelevant to the activity of most scientists.

However, Janik herself goes on to write that such enquiry is

> essential to the solving of a group of fundamental questions on the frontier of science and metaphysics which no amount of experiment at the phenomenal level will settle, including the nature of the basic constituents of the universe, the relation of God to the world, and the possibility of human freedom.

I agree that these are all topics of importance to Du Châtelet (and we will come to each in the following chapters). Janik's mistake, I think, is to assume that there are two projects here, with science as one project and providing a metaphysical foundation as the second. At the time Du Châtelet was writing, science and philosophy had yet to go their separate ways, and it was not at all clear that "physics" could proceed as an autonomous discipline. Where from a 20th- and 21st-century perspective we might be inclined to see two projects, Du Châtelet saw only one: the search for a causal understanding of the properties and behaviors of all bodies. This is what she means by "physics" (see earlier). What Du Châtelet shows, in attempting to carry out this project, is just how much metaphysics will be needed in order to see it through. I address this issue in the chapters that follow.

In translating "*Institutions de Physique*" as "*Foundations of Physics*", we therefore face problems. We must not read "foundations of physics" as "metaphysical foundations for what we would today call physics". By "Institutions", Du Châtelet does not mean "metaphysical foundations", though this forms a part of her discussion.[28] Rather, "foundations" more closely aligns with the usage by philosophers of physics today than that by post-Kantian metaphysicians, in her exploration of the conceptual foundations of physics. Her concerns include the epistemic status of theoretical claims, the logical consistency of theories, the presuppositions of theoretical claims, their ontological implications, and so forth. The alignment with conceptual foundations of physics in 21st-century philosophy is far from perfect, however. Most importantly, by using the word "physics" Du Châtelet is indicating that *causal* concerns are central to her project, and she is thereby contrasting herself with those who do "mechanics".[29]

Du Châtelet offers a striking image of her own view of her text in the Preface (2009, Preface, XI):

> Physics is an immense building that surpasses the powers of a single man. Some lay a stone there, while others build whole wings, but all must work on the solid foundations that have been laid for this edifice in the last century, by means of geometry and observations; there are still others who survey the plan of the building, and I am among these latter.

Du Châtelet's primary goal was not to add a stone here or there to the edifice of physics but to survey the plan of the whole.[30] It is in this sense that she is concerned with the *foundations* of physics.

Before concluding this section, I offer a brief digression. In thinking about how to frame Du Châtelet's project, it may be helpful to consider it in light of recent work on Kant's early philosophy. In the introduction to his monograph on the early writings of Kant, Martin Schönfeld writes (2000, p. 10), "The theme running through Kant's early philosophy was the thick cable of his precritical project, the reconciliation of natural science and metaphysics". We know that Kant was acquainted at least with Du Châtelet's work on *vis viva*, and there are sufficient similarities in their work such that Du Châtelet should be of crucial interest to anyone interested in the precritical Kant. The following quotation is taken from my notes for my opening remarks on the very first day of the Du Châtelet seminar that I taught at Notre Dame in 2014. It is an extract from Schönfeld, with comments inserted. The intention is by no means to criticize Schönfeld. On the contrary, his book plays a major role in making visible the philosophical problems associated with the overlap between matter theory, mechanics, physics, and metaphysics in the mid-18th century, and thereby Du Châtelet's contributions to philosophy. Here, it serves to highlight the interest of Du Châtelet's project:

> That the precritical project was an attempt to save the worsening relationship of natural science and metaphysics is also what makes Kant's early philosophy of nature unique. [That claim is false, this is exactly DC's project.] Kant recognized the rift between the scientific and metaphysical perspectives, and he wanted to do something about it. In his view, a reconciliation was needed because both perspectives are indispensable. [This is exactly what DC argues.] Natural science provides us with knowledge of the physical world; metaphysics provides us with answers to our questions about the intelligible framework of the physical world. This was Kant's precritical position. [And it was also DC's position.[31]]

In my notes, I go on to draw out more quotations from Schönfeld in order to emphasize the point that the central project of Kant's early work was that of Du Châtelet. For example, Schönfeld writes (p. 9), "Kant's early thought was guided by a vision of combining a modern mechanical model of physical nature with the metaphysical assumptions of a uniform structure of nature, of a purpose to the world, and of the possibility of freedom". As we shall see, this was Du Châtelet's vision, too.

I will argue that the problem of method lies at the heart of Du Châtelet's *Foundations* (see *Chapter 2*), and this Schönfeld sees in Kant too, writing (p. 14):

> Moreover, in contemplating the possibility of mediating between natural science and metaphysics, Kant and Lambert realized that the biggest obstacle was the absence of a coherent methodology that would do justice to either approach. Kant struggled with this challenge in the *Prize Essay* (1764); Lambert wrote two essays on the subject at the same time, which he subsequently incorporated into his *Neues Organon* (1764; "New Organon").

In short, when Schönfeld opens his book with the following description of Kant's precritical project, what we find is a description that fits Du Châtelet's earlier project equally well (2000, p. 3):

> I shall argue that the central theme in this period was Kant's struggle for a coherent philosophy of nature... This was Kant's precritical project, and it turned out to be the most ambitious venture of his life. He attempted to integrate Newtonian physics in a comprehensive and speculative framework that explained the macroscopic features of the universe as well as its microstructure, that accounted for its past as well as for its present, that permitted the copresence of rational freedom and deterministic lawfulness, and that illuminated the relation of God to the world.

It is indeed an ambitious venture. In my exposition of Du Châtelet's *Foundations of Physics*, I will not make reference to Kant for he comes later, and my goal is to read Du Châtelet's text on its own terms. The work of exploring the relationships between Du Châtelet and Kant remains to be done.[32]

To sum up: the "Received View" of Du Châtelet's *Foundations* consists of four main elements. First, the structure of the text is understood as consisting of two parts: the opening chapters on Leibnizian metaphysics and the later ones on Newtonian physics. Then, with the text understood in this way, two different approaches are offered. One approach characterizes the text as a translation and exposition of the views of Leibniz and Newton (among others). This approach has been decisively rejected by those who have argued for the second approach, according to which Du Châtelet was the first to attempt a synthesis of the views of Leibniz and Newton (among others). Nevertheless, even among those who adopt the synthesis view of the text, some commentators have fallen into a trap laid by the first approach: they have evaluated the text in terms of accuracy of exposition. More specifically, they have examined the extent to which Du Châtelet correctly presented the elements of Leibniz's metaphysics and Newton's physics on which she drew, and used this as the basis for evaluating the resulting synthesis. This evaluation is related to the fourth element of the "Received View", according to which Du Châtelet's primary goal in the *Foundations* is to provide a Leibnizian metaphysical foundation for Newtonian physics.

In my view, by contrast, the structure of the text is rather different, and this reflects the different interpretation that I offer of Du Châtelet's philosophical goals in the *Foundations* in terms of the problems of bodily action and of method (see Sections 1.1 and 1.3). These differences affect the way in which we evaluate the philosophy of the *Foundations*. In my opinion, the philosophical project of Du Châtelet's *Foundations* is an attempt to use resources from the best philosophy of the time, including resources that might fall under one of the general headings of Cartesian, Leibnizian, or Newtonian, in order to address what Du Châtelet took to be the most pressing philosophical issues of the time. If we can understand her philosophical objectives, we can then begin to understand the use to which she puts these resources and the philosophical moves that she makes. A careful reading of the philosophical content of the *Foundations* shows that she is active in her engagement with the resources on which she draws. She makes deliberate choices of which resources to use (and, equally importantly, which to omit) in light of her philosophical goals. She transforms those resources in the service of those goals. And she thereby arrives at new philosophical positions on a wide variety of issues and at a new overall philosophical system.

1.3 Overview of the Book

The approach to Du Châtelet's *Foundations of Physics* taken in this book is through the "problem of bodies", which I believe is the motivating philosophical problem lying at the heart of the text. The subject matter of physics in the early 18th century was the nature, properties, and behaviors of bodies, including their actions on one another. As Du Châtelet rightly saw, early 18th-century attempts to provide a theoretical account of bodies were riddled with difficulties, including disagreements and lack of clarity over the criteria that any successful account should have to meet. This lack of clarity makes it difficult to articulate the "problem of bodies" precisely, and I use the term as a label for the various problems related to the various attempts to provide some sort of account of the bodies that serve as the subject matter of physics.

Du Châtelet was particularly concerned with the question of bodily action: of how it is that bodies are capable of acting on one another. During the late 1730s, she became increasingly dissatisfied with the resources by which the Newtonians sought to answer this question. The *Foundations of Physics* offers her proposed solution. The means by which she arrives at this solution is a methodology that supplements an empirical approach championed by Newtonian philosophers with appeal to "principles of our knowledge", especially the principle of contradiction (PC) and the principle of sufficient reason (PSR). This supplement to the methodology was introduced in the revisions to the *Foundations* that took place immediately prior to publication. I believe that it was introduced precisely because it enabled Du Châtelet to address the problem of bodies, and that this is what drove the revisions to the text, including the introduction of non-extended simples as the constituents of extended bodies. Therefore, if we are to understand the role of these simples in Du Châtelet's philosophy, we must first understand the methodology that she sets out in the opening chapters of the *Foundations*.

My claim is that the problem of scientific methodology is a topic that both frames and unifies Du Châtelet's *Foundations of Physics*. The text begins from the problem of method and then applies that method in solving the problem of bodies, but this is not all. In the later chapters of the *Foundations*, Du Châtelet goes on to apply her method in addressing the most controversial issues in physics at the time concerning bodily action. These issues include Newtonian action-at-a-distance gravitation and the so-called "*vis viva* controversy". The upshot of my approach is a conclusive demonstration that Du Châtelet's primary motivation in introducing the Leibnizian and Wolffian elements into her philosophy was not to provide a metaphysical foundation for physics but to provide a method capable of resolving the biggest problems facing physics at the time. There is no question that, in resolving these problems by means of this method, she was led to introduce metaphysical commitments quite alien to her Newtonian contemporaries. Understanding why she came to do so deepens our understanding of her philosophy.

I make my case for this reading of Du Châtelet's *Foundations* as follows. In *Chapter 2*, I argue that the *Foundations* opens with the problem of method and I explain Du Châtelet's two-pronged methodology for science. I argue that the early chapters should be read as primarily methodological rather than metaphysical (notwithstanding the metaphysical commitments therein).[33] With this method in place, we can turn to Du Châtelet's treatment of the problem of bodies. In *Chapter 3*, I discuss her account of matter, body, and force. In this chapter, we see the way in which her methodological commitments (from *Chapter 2*) lead to the introduction of epistemological and metaphysical commitments that seem to take us a long way from Newtonian physics. A pressing difficulty of the time was how it is that there can be extended bodies at all. Du Châtelet addresses this question in Chapter 7 of the *Foundations* (Of the Elements of Matter), before turning to the problem that motivated her in the first place: bodily action. The problem of how it is that bodies are capable of acting on one another is addressed in Chapter 8 of the *Foundations* (Of the Nature of Bodies). Critically, Chapters 7 and 8 both depend on the new methodology, and especially on PC and PSR, in order to develop their solutions. Moreover, it is the employment of PC and PSR in their methodological role, in attempting to solve the problem of bodies, that leads to the introduction of the central metaphysical commitments of the *Foundations*; these include non-extended simples in her account of extended bodies, and her complex theory of forces (including the primitive force of non-extended simples) in her account of the agency of bodies. Points of interest in her account include a strong statement of "Laplacian determinism" more than half a century before Laplace, an account of the extension of bodies as arising from how we experience multiplicities, and a novel distinction between geometrical and physical space (or extension).

In *Chapter 4*, I discuss Du Châtelet's application of her method, and of her account of bodies, to controversial problems of bodily action in physics at the time. Ever since Descartes's *Principles of Philosophy* (first published in 1644), the problem of collisions had vexed philosophers across continental Europe and the British Isles: how are we to provide a causal explanation of the process of collisions by appeal to the nature of bodies? In *Chapter 4*, I show how Du Châtelet attempts to address this problem and discuss her commitment to mechanical explanation. I then present Du Châtelet's interventions in the two most controversial problems of bodily action of the day. The first concerns whether bodies can act on one another at a distance, as exemplified most successfully in Newtonian gravitation. The second is the so-called "*vis viva* controversy", concerning the proper measure of the force inherent in a moving body. As we will see, Du Châtelet invokes both her method and her account of bodies in seeking to resolve these issues.

In short, my claim is that the problem of bodies is a central and pressing problem for Du Châtelet; that she finds the Newtonian method to be impotent with respect to this problem; that shortly before publication she found a new method that had resources to make progress with the problem; and that the use of this method to address the problem is what led to the introduction of a range of new metaphysical commitments into the philosophy of the *Foundations.*

There is a great deal that I do not discuss, both within the text itself and in connection to Du Châtelet's wider philosophical corpus. References to papers treating some further aspects of the text in detail can be found in the notes. More work is needed to treat the variants of the text, Du Châtelet's sources, and the influence of her text on later work in philosophy. For a comprehensive view of Du Châtelet's philosophy, we await the forthcoming monograph by Detlefsen and Janiak.

Finally, I have tried to make the chapters of this book as self-standing as possible, so that they may be read independently of one another depending on interests and time. This inevitably results in a little repetition here and there, for which I hope I will be forgiven.

Notes

1 Du Châtelet wrote several other pieces of philosophical interest, including works on physical science (on light and fire, on optics, and the commentary on and translation of Newton's *Principia*) as well as addressing other topics, including ethics, morality, free will, the education of women, happiness, and writing a commentary on the Bible. See the edited collection by Hagengruber (2012a) for more details, including the bibliography by Rodrigues (2012). Ehrman (1986) provides an introduction to several of Du Châtelet's texts: the *Foundations*, her translation of and commentary on Newton's *Principia*, her translation of and commentary on Mandeville's *Fable of the Bees*, her *Grammaire raisonnée*, her commentary on the Bible (*Examen de la Genèse*), and her *Discours sur le bonheur.* See also Zinsser and Hayes (2006). For a list of Du Châtelet's known works, both published and unpublished, and the most up-to-date information on the availability of English translations, see Project Vox (https://projectvox.library.duke.edu).

2 A complete English translation of Du Châtelet's *Foundations of Physics* is available via three sources. See Zinsser and Bour (2009) for the Preface and Chapters 1, 2, 4, 6, 7 (partial), 11 (partial), and 21 (partial). See Patton (2014) for Chapter 9. For the remaining chapters and passages, as translated by Brading et al., see Du Châtelet (2018). Zinsser and Bour (2009) provide some information about the text, including extracts from relevant letters, and in her biography of Du Châtelet, Zinsser (2006, pp. 169–89) provides an overview of the background to, writing of, and content of the *Foundations.* For a review of the secondary literature on the *Foundations*, see Section 1.3.

3 Zinsser (2006, pp. 209–10) writes that Père Jacquier supervised the Italian translation, that he facilitated Du Châtelet's election to the Bologna Academy of Sciences, and that Laura Bassi used the *Foundations* in her lectures at the University of Bologna. More research is needed on the influence of the *Foundations* in Italy and also in Germany.

4 See Carboncini (1987) and Maglo (2008). I have also benefited from the unpublished work of Anne Seul on this issue, whose research on Du Châtelet's connection to the *Encyclopedia* through Samuel Formey was presented at a workshop at the University of Notre Dame (April 2015), an Émilie Du Châtelet workshop at the University of Oxford (May 2015), and at a conference on Émilie Du Châtelet at the University of Notre Dame (April 2018). Seul's presentations included the discovery of additional *Encyclopedia* entries originating in the *Foundations*. These additional entries were also discovered by Glenn Roe, who discusses Du Châtelet's presence in the *Encyclopedia* in his (2018).

5 See Voltaire (1738) for the English translation. The connection to Du Châtelet's decision to write the *Foundations* is discussed in, for example, Janik (1982), Zinsser (2006, pp. 145ff.), Hagengruber (2012b, p. 5), and Kawashima (2004).

6 For more details of this context, both intellectual and social, see Janik (1982), Hutton (2004), Fara (2004, pp. 88–105), Zinsser and Hayes (2006), and Zinsser's biography (2006). The late 1730s was a period of intense intellectual activity for Du Châtelet, during which several of her philosophical and scientific manuscripts were written, and during which she and Voltaire carried out their experiments relating to heat, light, and fire (see Janik, 1982, pp. 85–6 for a summary). For Du Châtelet's many other responsibilities and activities during this period, in her role as the Marquise Du Châtelet, see Zinsser (2006). Harth (1992) considers Du Châtelet's work from the point of view of the challenges she faced as a woman at the time.

7 See Iltis (1977, p. 30), Janik (1982, pp. 86–7), and Zinsser (2006, pp. 119–20, 145–7) for more details of Du Châtelet's reading in philosophy prior to writing the *Foundations*, and Janik (1982, p. 92) for a brief comment on Du Châtelet's annotations of Plato's *Dialogues*. Harth (1992, p. 198) highlights the importance of Descartes for the philosophical positions developed by Du Châtelet. For information about which Leibniz and Wolff sources were available to Du Châtelet, and when, see Winter (2012).

8 More work on the influence of Locke is needed. Examples of Du Châtelet's engagement with Locke include her rejection of "thinking matter", her concerns with free will and bodily action, and her discussion of time (see *Chapter 3*). See also Hagengruber (2012b, pp. 8–13). Du Châtelet's remark that Newton believed "with Mr. Locke that one can explain the creation of matter through Space, imagining that God would have rendered several regions of Space impenetrable" shows that her reading of Locke included the second edition of Coste's French translation of Locke's *Essay* for this is where we find the pertinent footnote. See Bennett and Remnant (1978).

9 See Wade (1969, pp. 277 and 285) and Zinsser (2006, p. 167).

10 For the evolution of the *Foundations*, and a comparison of the manuscript with the published version, see Barber (1967), Wade (1969, pp. 286–8), Janik (1982, pp. 99ff.), as well as Lascano (2011) for the chapter on God.

For discussion of changes relevant to our purposes here, see *Chapter 2*. I am indebted to the ongoing work of Lauren Montes to transcribe and annotate the manuscript.

11 See Zinsser (2006, pp. 170–71). Maupertuis began tutoring Du Châtelet in 1733, and by around the fall of 1734, she had turned to Clairaut for additional instruction in mathematics (see Zinsser 2006, pp. 66 and 72). Janik (1982, p. 95) connects Maupertuis's visit to Du Châtelet at Cirey in January 1739 with the removal of several mathematically challenging sections of the manuscript and the subsequent appointment of a mathematical tutor. Much to Du Châtelet's disappointment, she was unable to persuade Johann II Bernoulli to take this position, and König accepted instead.

12 On 27 February 1739, Du Châtelet wrote to Frederick the Great that "a student of Wolff" would be assisting her with her quest to master the geometry that she needed as "the key to all the doors", including to the causes of the elasticity of bodies (by which they rebound), for which she believed neither the Cartesians nor the Newtonians had a satisfactory explanation (doi:10.13051/ee:doc/voltfrVF0900256a1c).

13 For more details of the relevant evidence, see the recent secondary literature, including contributions to the Hagengruber (2012a) edited collection, Zinsser (2006), and references therein. I agree with Janik's (1982) assessment, also argued for by Wade (1969), that there is continuity in Du Châtelet's project from its inception through to its completion. See also Hutton (2004, p. 530).

14 See Kawashima (2004, pp. 50–51) for a brief discussion of Du Châtelet's intentions for the book as expressed in the frontispiece of the 1740 edition. For discussion of Du Châtelet's positioning of her text, see Terrall (1995, p. 293). See also the anonymous review of the *Foundations* in the *Journal des sçavans* (published in two parts, December 1740 and March 1741), which discusses Du Châtelet having written for her son (December 1740, pp. 737–55), and which also suggests that she is overly modest in the *Foundations* about its goals and achievements (p. 740).

15 Detlefsen (2014) compares the structure of Du Châtelet's text to Descartes's *Principles of Philosophy*. She argues that both begin with indubitable principles of knowledge from which they proceed to conclusions about the metaphysics of God, and that they both go on from there to further conclusions concerning the metaphysics of the world that, in turn, ground the later chapters on physics. She writes (2014, p. 6), "One interesting element of what could be characterized as her admiration of Descartes's method is that the *Foundations* fairly closely tracks the structure of Descartes's project in his *Principles of Philosophy*". I think that when comparing the structure of Du Châtelet's text with that of Descartes's *Principles*, we have to see this as mediated by the Newtonian textbooks, which her text so directly parallels. Of these textbooks, Keill's was the first, and Barfoot (1990, pp. 171–2) argues that Keill modeled the structure of his lectures on Rohault's (1671) Cartesian textbook, yielding an indirect connection to the structure of Du Châtelet's *Foundations*.

16 Many of these texts go on to consider other topics, including our planetary system, optics, heat and fire, and so forth. The "Avertissement" at the beginning of the *Foundations* describes what follows as "Book I", and "Book I" appears at the bottom of the first page of the Preface

(and elsewhere throughout the text), so like other Newtonians, it seems that she originally intended additional volumes covering other topics. In her letter to Frederick the Great of 25 April 1740, accompanying an advance copy of the *Foundations*, she wrote, "The work will have more volumes, for which printing has begun of only the first". In Chapter 15, Paragraph 351, she remarks that there will another book in which she will discuss "our planetary world". However, further volumes were never forthcoming, even though she wrote on optics, had published an essay on the nature of fire, and had originally intended a chapter on human liberty for the first volume (about which more later; see *Chapter 3*). Instead, she went on to translate and write a commentary on Newton's *Principia*, work that she completed only shortly before her untimely death.

17 Du Châtelet's motivations for introducing this material are addressed over the course of this book.

18 The view found in some of the literature, that Du Châtelet began with a text in Newtonian physics, to which she added a foundation of Leibnizian metaphysics in her last-minute revisions, is mistaken and also hazardous. For example, Erhman (1986, p. 47) writes that the shift to metaphysics includes Du Châtelet's discussions of the divisibility of matter and inertia, yet these are topics found in the standard Newtonian texts on physics. The distinction between physics and metaphysics, as it was understood at the time, must be handled with care; see the discussion that follows. For more on the "Received View" of the *Foundations* as a marriage of Leibnizian metaphysics and Newtonian physics, see Section 1.2.

19 Musschenbroek (1744) is the English translation of a revised version of his 1736 *Elementa Physicae*.

20 Detlefsen (2014, p. 3) attempts to distinguish physics from natural philosophy as follows:

> Du Châtelet's commitment to developing a natural philosophy (rather than writing simply about physics) required that she find an appropriate metaphysics to ground Newtonian physics, thus presenting a unified system to replace the (in her and others' view) defunct system of Descartes.

I disagree with this interpretation of the terms. I do not see the terms "natural philosophy" and "physics" being used in importantly different ways at the time, so as to indicate that the former but not the latter includes metaphysical foundations (the full title of Newton's *Principia* is, after all, the *Mathematical Principles of Natural Philosophy*). As noted, I think the distinction between physics and metaphysics lay then not where we might today assume. Finally, as I argue later, I do not believe that Du Châtelet's primary objective was to provide a metaphysical foundation for physics.

21 This approach to the text seems to go back to the very first reviewer, who divided the book into a first part concerning metaphysics and a second part concerning physics. The *Journal des sçavans* published this two-part review in December 1740 (pp. 737–55) and March 1741 (pp. 135–53). In the first part, the anonymous reviewer treats the Preface and first eight chapters of the *Foundations*, and he (for we may safely assume that the reviewer was male) explicitly labels them as the "metaphysical" chapters of the book. In the second part, the reviewer treats the remaining chapters

(Chapters 9–21), which he states have "physics" as their subject matter. I disagree with this characterization of the structure of the book (see Section 1.1). Barber (2006, p. 21) notes that the opening sections of the manuscript version of the *Foundations* concerned scientific method, but maintains that in the published version these are replaced by chapters on metaphysics (see also *Chapter 2*).

22 As noted, the English-language secondary literature on the philosophy of the *Foundations* is quite small. References to this literature can be found throughout this book, and see the bibliography by Rodrigues (2012). I have also benefited from unpublished manuscripts on Du Châtelet's philosophy by Karen Detlefsen, Jamee Elder, John Hanson, Anne-Lise Rey, Anne Seul, Monica Solomon, Jeremy Steeger, and Aaron Wells. In addition to the English-language scholarship, there is important scholarship on Du Châtelet's philosophy in French, especially by Anne-Lise Rey and Keiko Kawashima. More generally, there is a vast literature on Du Châtelet, but only a very small proportion of this concerns her work in philosophy and science, and even there, the primary focus is all too often on her work in relation to Voltaire. As Terrall (1995, p. 283) notes, Du Châtelet

> has gone down in history first of all as Voltaire's mistress and secondly as the only French woman of her time seriously to develop her talent for mathematics and physics. Until recently, her story has been told primarily in light of Voltaire's, as if he were what made her interesting.

Terrall discusses the social context within which Du Châtelet sought to pursue her scholarly interests and to receive recognition for her scholarly achievements. Wade (see especially his 1969, pp. 276–91) began the case for Du Châtelet's independence as a thinker, and more recent scholarship is vindicating and developing the case for treating her as a philosopher in her own right.

23 If one comes to Du Châtelet's *Foundations of Physics* as an Enlightenment scholar with a specialist interest in Voltaire, then the most striking contrast between Voltaire and Du Châtelet is surely her incorporation of Leibnizian principles, and one can see how the "Received View" would naturally arise. It has persisted, despite the cautionary remarks of the Voltaire scholar Ira Wade (1969), and continues to influence scholarship on the philosophy of Du Châtelet. Nevertheless, this framing of the text is inappropriate for philosophers because it does not treat the text on its own terms. Here, in this and the following chapters, I offer an extended argument for an alternative framing of the text.

24 Du Châtelet was familiar with Clarke's version of Rohault's text, to which he appended extensive pro-Newtonian footnotes (see Hutton, 2012).

25 See Stan (2018) for more on the importance of Wolff (rather than Leibniz) as a source for Du Châtelet. Suisky (2012) puts the importance of Du Châtelet's integrative project as follows (p. 117):

> The scientific life of scholars who were working in the first decades of the eighteenth century was dominated by the reception of the legacy of Descartes, Newton and Leibniz. Neither Newton nor Leibniz nor their contemporaries gave a review of the state of art in mathematics,

> physics and philosophy where they took a position which was beyond the quarrels between the different schools. Du Châtelet's intention was to make such a complete re-examination possible.

He compares Du Châtelet's *Foundations* to the later *Letters to a German Princess* of Euler and also suggests (pp. 114 and 117) that her presentation may have been of importance for Euler's own approach. Suisky argues that Du Châtelet, like Euler, attempted a "reception and reinterpretation" of Newtonian and Leibnizian principles in the context of a shared Cartesian legacy. He points to Johann I Bernoulli's attempt to draw on both Descartes and Newton in his *Essai d'une nouvelle Physique céleste* (1735) as a possible source of inspiration for Du Châtelet (p. 118).

26 Iltis (1977) concludes her section on Du Châtelet's "philosophy of nature" thus (p. 38):

> In evaluating Madame du Châtelet's philosophy of nature we may say that she did not claim originality for the philosophical ideas she expressed. She regarded herself as the disseminator and translator of the work of others. Philosophically her ideas were often confused and inconsistent. She did not grasp the subtleties of the full Leibnizian doctrine but this was due in part to the limited availability of texts, the failure of Leibniz himself to provide a systematic exposition of his own philosophy, and its dissemination through the secondhand accounts of Wolff and Koenig.

27 On 25 April 1740, Du Châtelet wrote to Frederick the Great about her forthcoming book, which she describes as a book in metaphysics. As Besterman writes in his notes on the letter, this rather misrepresents the content of her book. In reading her letter, we must take into account that Frederick had taken the trouble of supplying her with a French translation of materials from Wolff. She wrote that she had to write the book for her son because there was no complete book of physics in French, but that since she is

> persuaded that physics cannot proceed without metaphysics, on which it is founded, I wanted to provide him with an idea of the metaphysics of Mr. Leibniz, which I must admit is the only one that has satisfied me, even though I continue to have many doubts.
>
> (doi:10.13051/ee:doc/voltfrVF0910155a1c)

Even with the context taken into account, this letter clearly indicates a foundational role for metaphysics. I do not dispute this. My goal is to urge a broader conception of Du Châtelet's foundational project. Detlefsen (2014) writes that in the *Foundations* "Du Châtelet sought a metaphysical basis for the Newtonian physics she embraced upon rejecting Cartesianism" (p. 1) and that she "supplied the metaphysical basis for the Newtonian physics she had long accepted" (p. 2). In this, Detlefsen follows the "Received View", but she also indicates the possibility of a broader conception of that foundational project, beyond metaphysics, when she writes (p. 33), "The first ten chapters of the *Foundations* provide the epistemology, metaphysics and theorizing on scientific methodology that is meant to provide the foundations for the remaining chapters on Newtonian physics".

28 Harth (1992, p. 190) translates “Institutions” as “Lessons”, thereby avoiding unwanted connotations of “Foundations”. In the fall of 2014, Andrew Janiak and I were discussing the translation of the title, and I quipped that the only unproblematic word is “de” (of). I was wrong. Harth’s translation is “Lessons *in* Physics”.

29 See also Du Châtelet (2009, 2.27). The relationship between physics and mechanics in the 18th century is the subject of a collaborative research project between myself and Marius Stan.

30 It is notable that Du Châtelet’s characterization of science makes it a collaborative enterprise, involving many people, who play several different roles.

31 My views about this particular claim have become rather more nuanced since my first encounter with the *Foundations* in 2014, as what follows in this book makes clear.

32 Massimi and De Bianchi (2012) include a discussion of the role of Du Châtelet’s exchange with Mairan on the topic of *vis viva* (see *Chapter 4*) in Kant’s early matter theory.

33 A word of warning on terminology: Du Châtelet does not use the phrase “scientific method” or “scientific methodology”, and for her metaphysics is a “science”. I therefore use these phrases with deliberate anachronism for the meaning that they convey to philosophers today.

2 Method

2.1 The Problem of Method

We begin our philosophical journey by locating ourselves in France in the mid-1730s. The fourth edition of Newton's *Opticks* with the full set of *Queries* is brand new (it was published in 1730), and the third edition of Newton's *Principia* (published in 1726) is only a decade old. The standard French physics textbook, Rohault's Cartesian *Traité de physique* (published in 1671), predates the first edition of Newton's *Principia* (of 1687) and is out of date. Textbooks in Newtonian physics have been appearing in Latin (and in English), but there is nothing in French. In 1738, Voltaire published his *Éléments de la Philosophie de Newton*, explaining Newton's theories of light (from the *Opticks*) and gravitation (from the *Principia*). Émilie Du Châtelet, who worked with Voltaire on this text, knew that it did not fill the need for a French textbook on Newtonian physics; she decided that she would be the one to provide it.

For further instruction in Newtonian physics Voltaire, in his *Éléments*, sends the reader to Keill, 's Gravesande, Musschenbroek, and Pemberton (see Voltaire, 1738, p. 4). The textbooks of Keill (1702), 's Gravesande (1720), Pemberton (1728), and Musschenbroek (1734) all follow a common format, and Du Châtelet does the same: see Appendix 1 for the titles and chapter headings of these works.[1] What makes her text so philosophically groundbreaking is the content that goes into this format. Originally envisaged as a Newtonian textbook, Du Châtelet's *Foundations of Physics* breaks with her Newtonian predecessors, and it does so first on the issue of appropriate method for pursuing physics. Physics, in the early 18th century, was the study of bodies, including their "nature, attributes, properties, actions, passions, situation, order, powers, causes, effects, modes, magnitudes, origins" (Musschenbroek, 1744, p. 2), as we saw in *Chapter 1.*

Our question now is: by what method are we to pursue this project? As we will see, Du Châtelet found the methods offered by the Cartesians and the Newtonians inadequate. Instead, she offered a two-pronged methodology drawing on both empirical resources and *a priori* principles of knowledge. On the empirical side, she was the first to articulate a detailed hypothetico-deductive approach, emphasizing such familiar elements of this approach as falsificationism and the avoidance of *ad hoc* modifications in the face of apparent falsifications (see Section 2.3). For the principles of knowledge, she drew on resources from Leibniz including (most importantly) his principle of contradiction (PC) and principle of sufficient reason (PSR) (see Section 2.2).

In *Chapter 3*, we will consider the problems in physics that led Du Châtelet to conclude that the methods of the Cartesians and the Newtonians are inadequate for the purposes of physics. In this chapter, we look at the details of her proposed alternative methodology. My first task is to show that the question of method is indeed the central issue of the opening chapters of the *Foundations.* This is because, as we saw in *Chapter 1*, the "Received View" of the text is that the opening chapters primarily concern metaphysics. We will see in *Chapter 3* that these chapters do indeed contain significant metaphysical content. However, we cannot understand why that metaphysical content is there without first understanding Du Châtelet's primary philosophical purpose in these chapters. And this, as we will now see, was to address the problem of method.

Each of the Newtonian textbooks listed earlier begins with a discussion of method in philosophy: Keill begins with Lecture I, "Of the Method of Philosophizing"; 's Gravesande's Chapter 1 is entitled "Of the Scope of Natural Philosophy, and the Rules of Philosophizing"; Musschenbroek's Chapter 1 is "Concerning Philosophy, and the rules of Philosophizing"; and Pemberton's book begins with an "Introduction concerning Sir Isaac Newton's method of reasoning in philosophy". All of these opening chapters, except Keill's, contain Newton's rules of reasoning. Keill himself says nothing in Lecture I about Newton's rules, but in Lecture VIII his 16 Axioms and his discussion thereof contain within them Newton's first three rules. His text is particularly interesting because the original lectures of 1700 took place more than a decade before the second edition of the *Principia,* in which Newton published revised rules of reasoning and labeled them as such (instead of as "hypotheses").[2] In beginning his 1700 lectures with the question of method in philosophy, Keill picked up an issue that was controversial in the wake of the initial publication of the *Principia*, and central to the disputes over the status of Newton's claims concerning universal gravitation in Book 3 of the *Principia*. Moreover, by beginning with the "method of philosophizing", Keill seems to have set the model for later Newtonian texts.

Similarly, we know that Chapter 1 of the original manuscript version of Du Châtelet's *Foundations* contained Newton's third rule and therefore, we may suppose, at least his first three rules. From this we should conclude that, in the original version of her text, Du Châtelet began as her fellow Newtonians began: with a discussion of method, and one that included Newton's rules of reasoning. However, in the published version of the *Foundations*, Newton's rules of reasoning have disappeared, and an entirely new method takes center stage. She had examined the method of the Newtonians, and she had found it wanting.[3]

In the Preface to the *Foundations of Physics*, Du Châtelet begins by situating her text, first as a pedagogical text written for her son, second as a text directed toward the current state of knowledge in physics, and third in the context of the existing literature. She then turns our attention to the issue of method. Central to the Newtonians' position on method was their rejection of the Cartesian method, which they characterized as rampant system-building by means of speculative hypotheses and inadequately constrained by attention to the details of observation and experiment. The most famous formulation of this is Newton's own succinct rejection of hypotheses in the *Principia*: "hypotheses non fingo".[4] Pemberton discusses the hazards of hypothesis and conjecture, and of system-building, writing (1728, p. 3):

> The custom was to frame conjectures, and if upon comparing them with things, there appeared some kind of agreement, though very imperfect, it was held sufficient. Yet at the same time nothing less was undertaken than entire systems.

Musschenbroek's solution was to ban hypotheses (1744, p. 8):

> Therefore hypotheses are to be utterly banished from Physicks; for whatever is deduced from them must be uncertain, nor can be esteemed as demonstrated. And besides science will rather be oppressed than advanced by feigning hypotheses. Useless controversies will be raised, and the phenomena will be distorted, nay perhaps feigned, that the hypotheses may be defended and confirmed.[5]

As we will see, Du Châtelet proposed a different solution, arguing for the importance of hypotheses in scientific theorizing. Nevertheless, she recognized the epistemic risks of the conjectural method, writing (2009, Preface, VIII):[6]

> It is true that hypotheses become the poison of philosophy when they are made to pass for truth… an ingenious and bold hypothesis, which has some initial probability, leads human pride to believe it, the mind applauds itself for having found these subtle principles, and next uses all its sagacity to defend them. Most great men who have made systems provide us with examples of this failing. These are great ships carried by the currents; they make the most beautiful maneuvers in the world, but the current carries them away.

A shared goal among the Newtonians was to make progress in physics whilst avoiding error,[7] and the texts include discussions of our epistemological limitations and the appropriate standards of demonstration in natural philosophy. Keill, for example, writes (1720, p. 88):

> I would not have any one in physical Matters, insist so much on a rigid Method of Demonstration, as to expect the Principles of Demonstrations, that is, Axioms so clear and evident in themselves, as those that are delivered in the Elements of Geometry: for the Nature of the thing will not admit of such.

And similarly Pemberton states (1728, p. 19):

> The proofs in natural philosophy cannot be so absolutely conclusive, as in the mathematics. For the subjects of that science are purely the ideas of our own minds… they are themselves the arbitrary productions of our own thoughts, so that as the mind can have a full and adequate knowledge of its own ideas, the reasoning in geometry can be rendered perfect. But in natural knowledge the subject of our contemplations is without us, and not so compleatly to be known, therefore our method of arguing must fall a little short of absolute perfection.

Nevertheless, Newtonian philosophers sought a method that would yield results in which they could have confidence and which, in Newton's words, could "be considered either exactly or very nearly true notwithstanding any contrary hypotheses, until yet other phenomena make such propositions either more exact or liable to exceptions"

(Newton, 1999, p. 796). The problem facing Du Châtelet is apparent in Pemberton's text when he writes (1728, p. 5):

> the only method, that can afford us any prospect of success in this difficult work, is to make our enquiries with the utmost caution… for in this spacious field of nature, if once we forsake the true path, we shall immediately lose ourselves, and must for ever wander with uncertainty.

But proceeding with "utmost caution" is hardly a method. How are we to proceed? Here is Pemberton again (1728, pp. 19–20):

> It is only here required to steer a just course between the conjectural method of proceeding, against which I have so largely spoke, and demanding so rigorous a proof, as will reduce all philosophy to meer scepticism, and exclude all prospect of making any progress in the knowledge of nature.

According to Pemberton, and to Keill, 's Gravesande, and Musschenbroek too, we steer this "just course" by following Newton's rules as a guide for inductive reasoning. But Newton's brief statements are inadequate for this purpose. Du Châtelet removed his rules from her text and replaced them with her alternative method: a two-pronged approach designed to enable us to steer a just course between the Scylla of fictional conjecture and the Charybdis of skepticism. On the one hand, we are to use experience and experiments.[8] She writes (2009, Preface, IX):

> In all your studies, remember, my son, that experiment is the cane that nature gave us blind ones, to guide us in our research; with its help we will make good progress, but, if we cease to use it, we cannot help falling. It is experiment that teaches us about the physical characteristics of things and it is for our reason to use it and to deduce from it new knowledge and new enlightenment.

On the other hand, we must make use of Leibniz's PSR (2009, Preface, XII):

> it seems to me that with the principle of sufficient reason, he has provided a compass capable of leading us in the moving sands of this science.

By "this science" she is here referring to metaphysics, but in Chapter 1 (2009, 1.8), Du Châtelet is explicit that the PSR applies to the sciences in general (see later). The upshot is a method according to which theorizing is to be constrained from two directions: from the principles of our knowledge (which are the subject of Chapter 1 of the *Foundations*), and from experience and experiment (discussed in Chapter 4). The goal is theories of causes and of causal relations among things that we are justified in accepting as highly probable or even as true (see later).[9]

I have passed over much that is of importance in the Preface, but my point has been to emphasize that the focus of the Preface is scientific methodology.[10] If we take Du Châtelet's text at face value, then the Preface is telling us that her principal concern is appropriate methodology for achieving knowledge in physical science. We should read her text in this light.

The most well-known aspect of Du Châtelet's treatment of scientific methodology is her chapter on hypotheses (2009, Chapter 4). In this chapter, Du Châtelet argues that the Cartesians admit too many hypotheses, that the Newtonians admit too few, and that the appropriate approach is to admit hypotheses while adopting more stringent criteria for assessing them. On the Cartesians, she writes (2009, 4.55):

> Descartes, who had established much of his philosophy on hypotheses… gave the whole learned world a taste for hypotheses; and it was not long before these fell into fictions. Thus, the books of philosophy, which should have been collections of truths, were filled with fables and reveries.

And in the same paragraph, she chastises the Newtonians, too: "Mr. Newton, and above all his disciples, have fallen into the opposite excess…". Scientific theorizing, she argues, requires hypothesizing (2009, 4.57 and 4.58):

> If we take the trouble to study the way the most sublime discoveries were made, we will see that success came only after many unnecessary hypotheses… for hypotheses are often the only available means to discover new truths. … Hypotheses must then find a place in the sciences, since they promote the discovery of truth and offer new perspectives….

What is needed are much stronger criteria for assessing hypotheses, and it is here that she explicitly states her two-pronged methodology (2009, 4.61, emphasis added):

> Without doubt there are rules to follow and pitfalls to be avoided in hypotheses. ***The first is***, that it not be in contradiction with the

> principle of sufficient reason, nor with any principles that are the foundations of our knowledge. ***The second rule*** is to have certain knowledge of the facts that are within our reach, and to know all the circumstances attendant upon the phenomena we want to explain. This care must precede any hypothesis invented to explain it; for he who would hazard a hypothesis without this precaution would run the risk of seeing his explanation overthrown by new facts that he had neglected to find out about.

The remainder of Chapter 4 is devoted to exploring the second prong of her method (see Section 2.3), the first already having been discussed in Chapter 1 (see Section 2.2).[11]

From the point of view of methodology, the major revision to the *Foundations of Physics* between the manuscript and the published version is the introduction of the first prong of her method. The original Chapter 1, which included Newton's rules of reasoning (see earlier), was replaced prior to publication, and Leibniz's principles of contradiction and sufficient reason (PC and PSR) were introduced. The second prong of the methodology (concerning the role of experiments and experience) was already in place before the revisions: the partial manuscript of the *Foundations* includes the chapter on hypotheses, and in comparing the manuscript version with the published version we find that there are only very small changes, all but one of which is minor. The sole important change occurs in Paragraph 4.61, the published version of which was quoted earlier. Instead of the two-pronged methodology, we are offered only the second prong (see Paragraph LXI of the manuscript, available online via the Bibliothèque nationale de France):

> Without doubt there are rules to follow and pitfalls to be avoided in hypotheses, being sure of the facts, and of the circumstances attendant upon the phenomena we want to explain....

In short, the only significant change between the two versions is the explicit introduction of the first prong of her new methodology: PSR and PC as principles to constrain theorizing.[12]

2.2 The Principles of Our Knowledge

The first prong of the methodology is discussed in Chapter 1, which is entitled "Of the Principles of Our Knowledge". Throughout, Du Châtelet uses the phrase "our knowledge" (not simply "knowledge"), emphasizing that this is her concern. Moreover, that her concern is not only epistemology, but also methodology, becomes immediately apparent: having stated that all aspects of our knowledge are founded on self-evident principles, the very first conclusion that she draws is this (2009, 1.I):

> So, it is very important to be attentive to principles, and the manner in which truths result from them, if one does not want to go astray.

The problem to be solved is clear. *How* we are to proceed in order to achieve knowledge and avoid error: by what method?

In the next paragraph, Du Châtelet rejects Descartes's method, arguing that reliance on clear and distinct ideas allows us to be misled by our imaginations and leads to perpetual disputes. We need a method that enables consensus and, according to Du Châtelet, such a method demands that we reason from accepted principles, proceeding by demonstration so as not to be misled by our imaginations, and admitting only those conclusions that are agreed upon by all to follow from the principles (see 2009, 1.II).

For Du Châtelet, the two principles of our knowledge are PC and PSR. Du Châtelet explicitly asserts that she is following Leibniz in adopting these two principles. There is a great deal in her treatment of these principles that warrants discussion, including a comparison of the differences and similarities of her treatment to that of Leibniz, Wolff, and others.[13] Crucial for our understanding of Du Châtelet's use of PC and PSR are the roles that they play within her own project, and for this purpose the most important elements of her treatment are as follows.

For Du Châtelet, PC and PSR are independent principles, with PSR being "neither less fundamental nor less universal" than PC (see 2009, 1.8). She writes that PC concerns necessary truths, whereas PSR concerns contingent truths (on which more in what follows).

Du Châtelet uses PC to distinguish between the impossible (that which implies a contradiction) and the possible (that which does not).[14] She writes that PC is "the foundation of all certainty in human knowledge", and moreover that without it there could be no knowledge, for without it "every thing could be, or not be, according to the fantasy of each person". Du Châtelet's discussion of PC (2009, 1.4–7) culminates with her emphasis on its methodological importance, in Paragraph 6 of Chapter 1. Here, she writes:

> A very important rule results from the definition of the impossible that I have just given you; it is that when we advance that a thing is impossible, we are required to show that the same thing

> is simultaneously asserted and denied... This rule would avoid a great many disputes, if it were followed....
>
> One should be just as cautious when maintaining that a thing is possible; for one must be in a position to show that the idea is free of contradiction.

Thus, to adopt PC as a principle of our knowledge is to adopt a method of reasoning concerning that which is possible and impossible.

The truths of geometry, Du Châtelet says, are necessary truths for which we need only PC. PC by itself, however, is insufficient for us to determine all the truths, since not all truths are necessary truths. Within the domain of that which is possible (as determined by PC) lie those things whose truth or falsehood is not necessary, but merely contingent. This is where PSR comes in: it is a tool for arriving at contingent truths. Du Châtelet writes that "when it is possible for a thing to be in several states" (2009, 1.8), PC is insufficient for determining its actual state, and PSR is required: we must provide a reason for why a thing is in one of its possible states rather than another:

> Thus, for example, I can be sitting, lying down, or standing, all these determinations of my situation are equally possible, but when I am standing, there must be a sufficient reason why I am standing and not sitting or lying down.

Du Châtelet provides a more detailed explanation of her account of the essential properties of a thing, its attributes, and its modes in Chapter 3. For our purposes here, the crucial point is that PSR is used to distinguish the actual from the merely possible.

To say that PSR is a tool for arriving at contingent truths underplays its significance for Du Châtelet. Its methodological role is intimately tied to its epistemological and metaphysical status in Du Châtelet's philosophy. This arises from Du Châtelet's view that PSR must be presupposed in order for knowledge of contingent truths to be possible at all, because without PSR no reasoning from effects to causes is possible.

Du Châtelet's primary target here is the Newtonians, with their reliance on inductive reasoning. Famously, Newton had advocated inductive practices, adding the fourth rule of reasoning in the third edition of the *Principia*, published in 1726 (Newton, 1999, p. 796):

> In experimental philosophy, propositions gathered from phenomena by induction should be considered either exactly or very

> nearly true notwithstanding any contrary hypotheses, until yet other phenomena make such propositions either more exact or liable to exceptions.

But it is only recently that we have begun to understand the import of this rule, and its interpretation remains controversial among Newton scholars.[15] At the time Du Châtelet was writing, the defense of induction in Newtonian texts was weak. Pemberton (1728) urges cautious use of induction as the only means by which to steer a course between wild conjecture and outright skepticism when it comes to the knowledge of the qualities of bodies and "the causes of things" (1728, p. 2), but offers no further justification. 's Gravesande (1720, Preface) argues that in physics we should follow inductive practices, just as we do in ordinary life, for

> unless we account those things as generally true, which every where appear as such, where we can make any Experiments; and that like Effects be supposed to arise from a like Cause, who can be able to live one Moment of Time In Ease?

He gives multiple examples of the inductive practices that underlie our everyday lives, writing ('s Gravesande 1720, Preface):

> I have used such and such Food for several Years, I will also take it to-day without any Fear.
>
> When I see Hemlock I conclude there is Poison in it, tho' I have made myself no Experiment about the Hemlock, which I see.

He then concludes:

> All these Reasonings are founded upon Analogy: And it is not to be doubted but we are put under the Necessity of Reasoning by Analogy, by the Creator of all Things. This therefore is a proper Foundation of Reasoning.

By the time Du Châtelet was finalizing her text, the first two books of Hume's *Treatise* (1739) had recently appeared, though they had yet to make an impact.[16] For those who sought to use empirical methods to arrive at secure and certain knowledge of the causes of phenomena, the lack of an adequate defense for inductive reasoning was there for all to see, and it is likely that Du Châtelet found the discussion of the epistemological basis for inductive reasoning in Newtonian texts unsatisfactory.

Du Châtelet makes a series of four points that support acceptance of PSR as a necessary prerequisite for the very possibility of knowledge. Her first point is as follows (2009, 1.8):

> If we tried to deny this great principle, we would fall into strange contradictions. For as soon as one accepts that something may happen without sufficient reason, one cannot be sure of anything, for example, that a thing is the same as it was the moment before, since this thing could change at any moment into another of a different kind; thus truths, for us, would exist only for an instant.

Du Châtelet was clear about the epistemological challenge: something is needed to tie the events together from moment to moment, for without this we can make no inferences beyond the particular moment in which we find ourselves. She gives an example in support of this point (2009, 1.8):

> For example, I declare that all is still in my room in the state in which I left it, because I am certain that no one has entered since I left; but if the principle of sufficient reason does not apply, my certainty becomes a chimera, since everything could have been thrown into confusion in my room, without anyone having entered who was able to turn it upside down.

Where Hume, according to many interpretations, found nothing to tie events together, Du Châtelet responded to Pemberton's plea that we avoid outright skepticism by accepting PSR as a prerequisite for the possibility of knowledge: it is the means by which we tie the succession of events together, such that we can reason from a state of affairs at one time to states of affairs at other times.[17] Without this, no reasoning about causes and effects, and therefore no inductive reasoning, is possible.

Du Châtelet's second point is that PSR is also required in order for us to make comparisons, including the comparisons involved in making measurements. Comparisons and measurements involve equality and inequality, identity and difference. Du Châtelet's claim is that without PSR, there can be no measurements, and empirical science would not be possible. The example Du Châtelet gives to make this point concerns the weight of objects (2009, 1.8):

> Thus, for example, if I have a ball made out of stone, and a ball of lead, and I am able to put the one in the place of the other in a basin of a pair of scales without the balance changing, I say that the

> weight of these balls in identical, that it is the same, and that they are identical in terms of weight. If something could happen without a sufficient reason, I would be unable to state that the weight of the balls is identical, at the very instant when I find that it is identical, since a change could happen in one and not the other for no reason at all; and consequently, their weights would no longer be identical....

The choice of weight is apt. Newton's *Principia* is all about gravity, and the argument of the *Principia* depends utterly on measurements of weight, from weighing terrestrial objects to "weighing" the Moon. Du Châtelet's position, implied in her reasoning here, is that unless we presuppose PSR the reasoning on which the argument of the *Principia* rests does not go through. Once again, the general point is that our reasoning about causes and effects is made possible by PSR.

Du Châtelet's third point about PSR is framed as a defense against atheism. Du Châtelet argues that rejection of PSR opens a route to atheism. Her argument is brief (2009, 1.8):

> Without the principle of sufficient reason, one would no longer be able to say that this universe, whose parts are so interconnected, could only be produced by a supreme wisdom, for if there can be effects without sufficient reason, all might have been produced by accident, that is to say, by nothing.

Here, Du Châtelet is again implicitly taking on the Newtonians who, like Newton himself, were partial to arguments from design.[18] Du Châtelet's claim is that without PSR, we have no warrant for inferring from what might look to us as though it is the product of a designer to the existence of such a designer, for we have no means of inferring from effect to cause.

Finally, Du Châtelet's fourth point about the need for PSR begins with the claim that it is through PSR that we escape the clutches of Descartes's dreaming argument (and the associated skepticism). PSR is the tool by which we are able to distinguish a dreamt world from waking reality: sequences of events in waking reality satisfy PSR, while those in a dreamt world do not do so (at least not for long enough). In the same way, PSR is the tool by which we are able to distinguish fable and fiction from reality. Here, Du Châtelet's target is the Cartesians, whose method has led them to produce books "filled with fables and reveries" (2009, 4.55). It is by means of PSR, according to Du Châtelet, that we can constrain our theorizing so as to fill our books of philosophy with truths instead of fantasies.[19]

The preceding considerations enable us to see the epistemic and metaphysical significance of PSR. For Du Châtelet, PSR is first of all an epistemic principle whose satisfaction allows us to reason from causes to effects and *vice versa*. In presupposing it, we presuppose that the universe cooperates: that metaphysically it is not capricious, and that it is therefore intelligible (at least to some extent) in terms of causes and effects. Without this, knowledge is not possible. For Du Châtelet, this metaphysical face of PSR is guaranteed by God (2009, Chapter 2). With this epistemic and metaphysical background in place, we are justified in using PSR as a tool for arriving at contingent truths, and we therefore impose PSR on our scientific theorizing as a methodological requirement. When it comes to contingent truths, PSR must be our guide.

Throughout the remainder of the discussion, Du Châtelet's main emphasis is on the methodological importance of PSR for scientific theorizing (2009, 1.8):

> for the source of the majority of false reasoning is forgetting sufficient reason; and you will soon see that this principle is the only thread that could guide us in these labyrinths of error the human mind has built for itself in order to have the pleasure of going astray.
>
> So we should accept nothing that violates this fundamental axiom; it keeps a tight rein on the imagination, which often falls into error as soon as it is not restrained by the rules of strict reasoning.

Moreover, within that methodological role for PSR, causes fall under the broader umbrella of reasons.[20] She writes (1.10), "a cause is good only insofar as it satisfies the principle of sufficient reason". This causal understanding of PSR is related to her version of the law of continuity, which she asserts as a corollary to PSR (1.13), and which plays a crucial role in her arguments throughout the *Foundations*. For Du Châtelet, unlike for Leibniz, the law of continuity follows from PSR (2009, 1.13):

> [E]ach state in which a being finds itself must have its sufficient reason why this being is in this state rather than any other, and this reason can only be found in the preceding state. Therefore this antecedent state contained something which gave birth to the current state that followed it, so that these two states are so completely

> interconnected it is impossible to put another state between the two. For if there was a state possible between the current state and that which immediately preceded it, the nature of the being would have left the first state without yet being determined by the second to abandon the first. Thus, there would be no sufficient reason why it should pass to this state rather than to any other possible state. Thus no being passes from one state to another without passing through the intermediate states.

Du Châtelet uses geometry to illustrate her conception of the law of continuity, arguing that it is satisfied only by smooth curves, and that all curves in nature are therefore smooth (including the path of a beam of light reflected at a mirror). Du Châtelet relates the smoothness of a curve to its having been produced by a single law (or function), whereas a curve with a discontinuity, she suggests, requires a second law, one for each part of the curve. As a result, such a curve is not a "true whole". The mathematics of continuity will make her claim problematic, but the philosophical point of interest is her position that a "true whole" satisfies the law of continuity and is produced by a single law.[21] For our purposes, the point is that the law of continuity, as a corollary of PSR, is central to Du Châtelet's methodology. Already in Chapter 1, Du Châtelet uses the law of continuity as a constraint on physical theorizing in order to argue that all changes in motion are continuous (1.14), that there are no perfectly hard bodies (1.15), and that Descartes's rules of collision are erroneous (1.17).

For our purposes, then, the key point is this: PSR is introduced in the *Foundations* as a principle whose acceptance is required in order to underwrite the very possibility of scientific theorizing, and whose implementation within theorizing is required as a component of Du Châtelet's methodology. PSR is *introduced* as a guide to our reasoning and is *used in practice* in Chapter 1 (and throughout the *Foundations*) as just such a guide.[22]

With this first prong of her method in place, Du Châtelet is able to tackle two other topics prominent in the opening chapter of each of the Newtonian texts mentioned earlier, and which relate to the goals, limits, and methods of our reasoning in physics: God (Chapter 2)[23] and our knowledge of the essences of things (Chapter 3). In *The Leibniz-Clarke Correspondence* (1998), one theme of contention is the clash between Leibniz's rationalist conception of God and Clarke's voluntarist conception. A shared theme in Newtonian textbooks is epistemic modesty and caution, and in the epistemology of both 's Gravesande and Musschenbroek, this is connected to a voluntarist conception of God.

The reason we must adopt an empirical approach to knowledge of the nature of bodies and the laws of nature, with all the fallibility involved in the associated inductive practices, is that they depend on the arbitrary will of God and therefore cannot be discovered by reason. Though physics is the search for causal explanations of the behaviors of bodies, the causal chain ultimately traces back to the will of God, and (writes Musschenbroek, 1744, p. 8) "when we arrive at the ultimate cause, which depends only on the will and power of God, we shall not perceive any clear connection between the cause and the effect". Moreover, Musschenbroek goes on, since "we have no rule or criterion certainly to know, when we have arrived at the ultimate corporeal or natural cause of things", we can never know whether we have arrived at the end point of our enquiries or whether further investigation into natural causes is appropriate. Prior to the *Foundations of Physics*, Du Châtelet had been willing to allow that explanations in physics may come to an end in the arbitrary will of God.[24] By the time of the revised and published version, she has decided instead to adopt a rationalist conception, using PC and PSR to reason about the nature and properties of God, and about the status of the essences of things.[25] All that is possible, including the essences of things, depends on PC, and not on the arbitrary will of God. It is therefore inadmissible in physics to appeal to the will of God when theorizing about how something is possible (2009, 3.49):

> Thus, one must admit nothing as true in Philosophy, when one can give no other reason of the possibility except for God's will, for this will does not make one understand how a thing is possible.

The actuality of this world, out of all possible worlds, depends on the will of God acting in accordance with his intelligence and wisdom.[26] His wisdom is manifest in the evidence of design in the world, and while Newton had made an argument from design in the *Principia*, Du Châtelet goes further, rejecting Musschenbroek's banishment of final causes from physics, and siding with Maupertuis in suggesting that final causes lie within the domain of physics, and that appeal to final causes is fruitful in physics (2009, 2.27). As noted, for Du Châtelet any such arguments from design, or involving final causes, rely on the presupposition of PSR.

In Chapter 3, we learn that the essence of a thing does not fully determine the ways in which it exists: it determines its attributes but not its modes.[27] While God's will is what gives actuality to the totality of this world, what gives actuality to the particular ways of being of a thing, out of the many possible ways, is its prior state and the prior state of those things with which it causally interacts. There is to be no appeal in physics to the will of God here either. PSR constrains the determination, via the law of continuity (which follows from PSR, according to Du Châtelet, as we have seen). This provides a reason for adopting PSR as a useful tool by which to constrain theorizing: metaphysically, the world satisfies PSR.

These last few paragraphs have been a hasty and unsatisfactory review of some of the themes of Chapters 2 and 3. A much fuller investigation of the interrelationships between Chapters 1–3 and the later chapters of the *Foundations* is needed, but lies beyond the scope of our enquiries here.[28] For our purposes, the take-home message is that Chapters 1–3 concern the constraints on theorizing in physics that arise from *a priori* considerations, and in particular from the adoption of PC and PSR. It is through this lens that they should, in the first instance, be read.

2.3 Observation and Experiment as a Constraint on Hypotheses

Perhaps the most well-known chapter of Du Châtelet's *Foundations* is Chapter 4, "On Hypotheses". Du Châtelet opens the chapter by explaining why hypotheses are needed in the sciences (see earlier), before explaining the second prong of her method: the use of empirical resources to constrain physical theorizing. This chapter of the *Foundations* was reproduced almost verbatim in the highly influential *Encyclopedia* of Diderot and d'Alembert, and while the influence of that entry has yet to be traced in detail, there can be no doubt of the importance of this chapter for the history of philosophy of science.[29,30]

Chapter 4 begins with a paragraph on the epistemic status of explanations in the sciences, with respect to the goal of finding the "true causes of natural effects and of the phenomena we observe".[31] Du Châtelet argues that there is an epistemic gap between true causes and both (a) the principles of our knowledge, and (b) experiments and experience, such that instead of requiring certainty we will have to accept explanations that are merely probable. We will need such explanations for two reasons: first, for their practical use, and second, because such explanations are helpful on the path to truth. Du Châtelet does not offer an account of what she means by "probable", or give any indication as to whether she is familiar with the mathematical approaches to probability dating from late in the previous century, but she does explain the conditions under which we are to find an explanation probable, and therefore to accept it, as we shall see.

Du Châtelet states that only those who are able to demonstrate the true causes of the phenomena have no need of hypotheses.[32] Since we are not in this epistemic position, we instead need to proceed in uncertain steps: we "can arrive at the truth only by crawling from probability to probability" (2009, 4.55).[33] Moreover, such progress will be a collaborative enterprise. In the Preface, Du Châtelet offered a collaborative view of science (as noted in *Chapter 1*), and this picture is further developed in Chapter 4, where Du Châtelet points out that making progress toward truth in a science involves making mistakes, especially in the early days, and that without these early errors by some, no progress by others would have been possible (see 2009, 4.54). This is why hypotheses are necessary in the sciences: we cannot know at the outset which hypotheses will lead us to truth, but it is only by making and evaluating such hypotheses that we can find our way there at all. Moreover, since this is a long and arduous process, it will involve the work of many (see Du Châtelet, 2009, 4.57). The example that Du Châtelet gives is astronomy, where the Ptolemaic geocentric hypothesis eventually led, through the investigation of the difficulties to which it gave rise, to the formulation of a new hypothesis, the Copernican heliocentric hypothesis (2009, 4.57).[34] Her conclusion is as follows (2009, 4.58):

> Hypotheses must then find a place in the science, since they promote the discovery of truth and offer new perspectives.

But, as we have already noted, not all hypothesizing is useful in science. Singled out for criticism in this regard are the Cartesians. As we saw, Du Châtelet followed a familiar line of argument from early 18th-century Newtonian philosophers, a theme that was already present in Newton's *Principia*. It was first eloquently expressed in print by Roger Cotes in his preface to the 1713 second edition of Newton's *Principia* (Newton, 1999, p. 386):

> Those who take the foundation of their speculations from hypotheses, even if they then proceed most rigorously according to mechanical laws, are merely putting together a romance, elegant perhaps and charming, but nevertheless a romance.

Du Châtelet echoes this in her criticism of the Cartesian use of hypotheses quoted earlier (2009, 4.55).

The problem with the Cartesian use of hypotheses was that qualitative consistency with observed phenomena was deemed sufficient (though not even always necessary) for explanatory success, so long as the explanation was consistent with the *a priori* principles of Cartesian philosophy. This is made clear by Descartes himself at the end of his *Principles of Philosophy* (see Part IV, Paragraphs 203–4). Such a method brings with it an immense problem of underdetermination, since *any* theory of microscopic behavior consistent with the *a priori* principles of Cartesian philosophy, and qualitatively consistent with our experience, counts as a successful explanation, and there is no way to distinguish between different explanations. For those seeking *the true* causes of the phenomena, this was far too weak a constraint on theorizing. Huygens (in 1690) sought to strengthen the role of experiment in the assessment of hypotheses, arguing for a hypothetico-deductive approach by which to attain "a degree of probability which very often is scarcely less than complete proof" (see Huygens, 2009, pp. 164–6). Du Châtelet went further, offering a two-pronged methodology including a rich treatment of how to use empirical resources in scientific theorizing. By means of the principles of our knowledge, on the one hand, and a strengthened use of empirical resources, on the other, Du Châtelet sought to rein in the use of speculative fictions associated with Cartesian physical theorizing while maintaining the necessary and useful role of hypothesizing in science.

Like the Cartesians, Du Châtelet accepted that the gap between fundamental principles and observable phenomena is so large that deduction from the principles will not be practicable. Like the Newtonians, she sought nevertheless to overcome the rampant underdetermination permitted by the Cartesian method, and to arrive at "probable" reasons for the observable phenomena (2009, 4.53 and 4.62). However, her method for using empirical resources is markedly different from the discussions of induction offered by Musschenbroek and by Pemberton, which offer little more than enumerative induction to yield conclusions of constant conjunction, and goes beyond the treatment offered by 's Gravesande.[35] The main features of Du Châtelet's account of how to use empirical resources in theorizing can be summarized as follows.

Du Châtelet begins by emphasizing that acceptance of a hypothesis depends on *all* its consequences agreeing with observations. We must explore the empirical consequences of a theory, and not merely seek consistency with prior observations (2009, 4.58 and 4.66). Moreover, we must seek *novel* predictions: we must explore the consequences of the hypothesis for phenomena other than those for which it was originally constructed. The breadth of success of a theory, as the new predictions made by the theory are confirmed by observation, will increase the probability of the hypothesis. This is how we assess how probable our explanations are, and thereby move as much toward demonstration as our epistemic situation allows: if "all the consequences drawn from it agree with the observations, its probability grows to such a point that we cannot refuse our assent to it, and that is almost equivalent to a demonstration". This statement echoes both Descartes's "moral certainty" (see his *Principles*, Part IV, Para. 205) and Huygens's statement quoted earlier.

However, in committing ourselves to a hypothesis we must be careful: we must put into the conclusion only that which must be there, and not give our assent to aspects of the hypothesis that play no role in its observational success (2009, 4.65). In other words, confirmation is *selective*.

Du Châtelet also emphasizes that there is an asymmetry between acceptance and rejection of a hypothesis in relation to empirical evidence, writing (2009, 4.64), "One experiment is not enough for a hypothesis to be accepted, but a single one suffices to reject it when it is contrary to it". She further develops her discussion of falsification, to point out that falsification too should be *selective*, so that we reject only that part of the hypothesis that is responsible for the falsification (2009, 4.65 and 4.66).

Finally, she insists that there be no *ad hoc* modifications, as a condition on a good hypothesis (2009, 4.69):

> it is necessary… that the phenomenon result necessarily, and without the obligation to make new suppositions … When the necessary consequences do not follow from it, and to explain the phenomenon, a new hypothesis must be created in order to use the first, this hypothesis is only a fiction unworthy of a philosopher.

Clearly, this is a rich treatment of the role of empirical investigation in relation to hypotheses in scientific theorizing, and it deserves considerably more attention in its own right.[36] The most striking thing about the account is, perhaps, how contemporary it sounds. We recognize all the elements of her account as present in today's discussions of scientific method in the philosophy of science literature.

For our purposes, the crucial points are that in Chapter 4 Du Châtelet restates her position that scientific theorizing is to be constrained from two directions – the principles of our knowledge, and experience and experiment – and offers a detailed account of how to use empirical evidence as a constraint on scientific theorizing. This is the method she will use in tackling open problems in the physics of her day (see *Chapter 4*).

2.4 Metaphysics and Method

I have suggested that Chapters 1–4 of Du Châtelet's *Foundations of Physics* should be read as primarily concerned with method. More precisely, Du Châtelet is concerned with the epistemological question of the status and limits of our knowledge of the world, and consequently with the methodological question of *how* it is that we can achieve knowledge in physics. However, as we saw in *Chapter 1*, the "Received View" of this text is rather different, emphasizing instead the metaphysics in the early chapters. Janik (1982, p. 93) writes:

> Studying the *Institutions* makes it clear that one of her basic convictions, from the beginning of its composition, was that science, whether Newtonian or otherwise, was dangerously incomplete without some kind of metaphysical foundation....
>
> She thus developed a plan for a different kind of introduction to the new physics, to consist of an enquiry into the fundamental properties of matter, the role of hypotheses, the nature of explanation, the function of God in the universe, the possibility of free will in a mechanistic world, and other such metaphysical problems. The structure of the *Institutions*, in which these metaphysical chapters precede the exposition of physical theory, indicates that by 1738 she had already come to hold that only in the context of such an enquiry could the significance of Newton's discoveries, and the value and limitations of his method, be properly expounded.

If we take this approach to the text, then we begin by observing that there are two halves to the text, a "Leibnizian metaphysics" half and a "Newtonian physics" half, and we seek to address the problem of whether and how, if at all, the two halves fit together. I think that reading the text as consisting of two halves, and reading the first half as primarily concerned with metaphysics, and with the provision of a metaphysical foundation for physics, is a mistake. It is true that the early chapters contain significant amounts of metaphysics, but I do not believe that this is because Du Châtelet was seeking a *metaphysical* foundation for the physics of the *Foundations*. Rather, her concerns over epistemology and methodology are what drive the introduction of the metaphysical content of the *Foundations*.

Indeed, Janik herself concludes (contrary to Barber (1967), but rightly I believe) that

> the 1738 version raised the same questions about methodology, including the use of hypotheses, in science; the existence and nature of God, the external world, and of human freedom; the properties of the ultimate constituents of the natural world; and the relation between our concepts (such as those of motion and force) and reality

and that in the published version "attention is still focused on the same *kinds* of question – in particular, the points of contact between physics and metaphysics, and the nature of human conceptualization of experience" (Janik, 1982, pp. 100–1). Like Hutton (2004), I see continuity in the revisions to the text, and I see Du Châtelet's concerns with method as crucial.

The case for my reading rests on both internal and contextual considerations. Internally, Du Châtelet herself seems to me to be quite explicit that her concern is with the problem of method: of *how* we are to arrive at true (or probable) theories in physics. Contextually, my claim is that Du Châtelet's text is best read against the backdrop of the Newtonian textbooks that she and Voltaire read in preparing his *Éléments de la philosophie de Newton* (1738). She is explicit in her Preface (see Paragraph III) that her goal was to produce the first up-to-date and complete physics textbook in French, replacing the outdated Rohault. My suggestion is that she took the Newtonian textbooks as her model, and that we should therefore expect her opening chapter (at least) to concern method, as the internal evidence indicates. If my suggestion is right, we should also expect her ensuing chapters to follow the same model, and this is indeed the case. Following the discussion of method, all the texts (except Pemberton) move immediately to the consideration of body, and there is, not surprisingly, considerable overlap in the topics covered thereafter, including body as extended and its divisibility; motion, place, space, and time; laws of motion or of nature; forces; rules of collision; terrestrial gravitational phenomena, including inclined planes, projectiles, and pendulums; the elasticity of bodies; and celestial gravitational phenomena. Du Châtelet's text follows the same pattern (see Appendix 1 for a list of chapter headings in these texts).

Some important conclusions follow from this analysis, concerning how we should approach the text. The first concerns our evaluation of the metaphysics. I do not deny that there is a great deal of "Leibnizian" metaphysics in several of the early chapters. However, I think that before examining and assessing Du Châtelet's metaphysics, we should be clear about the goals with which she introduced this material into her philosophy. The main quotation from Du Châtelet supporting a foundational project is from Du Châtelet's letter to Maupertuis of

23 October 1734 (my translation): "I divide my time between the builders and Mr. Locke, for I seek the foundation of things for one just as for the other". I do not dispute that she was seeking an adequate foundation for physics, nor that this foundation involves metaphysical commitments. My claim is that the primary engine of her foundational enquiry is the search for an appropriate methodology by which to achieve an epistemically secure (i.e. appropriately founded) physics. This approach to the text helps to explain the limits of the metaphysical content: as a metaphysical *foundation*, the metaphysics that she offers seems incomplete; but when viewed as a presentation of the least amount of metaphysics necessary for consistency with the epistemological and methodological constraints of her project, the limits of her metaphysics make more sense.

The second concerns the unity or disunity of the text. According to the approach I am advocating here, we should begin by assuming that Du Châtelet's text follows the standard Newtonian pattern outlined earlier, beginning with a discussion of method, moving into the topics of body, space, time, force, etc., and then moving on to particular cases concerning terrestrial gravitational phenomena and so forth (as listed earlier). As such, her text was, at least in the original manuscript version, a unified whole (to the extent that other similar Newtonian texts can be considered unified). Rather than following the "Received View", according to which the problem is whether and how there is any unity between the "two halves" of the text, our question now becomes: how, why, and to what extent (if at all) does the text become *dis*unified? How we approach the text has thereby been transformed. We begin by reading the opening chapters as primarily methodological, rather than as primarily metaphysical, and we seek internal philosophical reasons for Du Châtelet's dramatic rewriting of those chapters just as the text was going to press. Our expectation, on this approach, is that the revisions will concern method. It seems evident that these changes concerning method had associated with them the introduction of Leibnizian themes into the early chapters of the book. It might turn out that her method leads her to introduce tensions into her project that result in a disunity of the text. And it might be right that this is due to a tension between "Leibnizian" elements and "Newtonian" elements of that text. But, in contrast to the "Received View", we will not, and must not, start out by assuming that this is the case.

If the changes resulted in tensions in the text, then we will not be able to understand these unless we have a correct understanding of what drove the changes that Du Châtelet made right at the moment of publication. We take a look at this issue in *Chapter 3*.

Notes

1 For further discussion of Du Châtelet's reading during the 1730s, see Zinsser (2006, pp. 145ff). Keill (1720) is an English translation of lectures given by Keill in Oxford in 1700, first published in Latin (1702 and later editions). 's Gravesande (1720) is an English translation by John Keill of the original Latin work by 's Gravesande, which also appeared in 1720 (as did a competing English translation by Desaguliers). Musschenbroek published multiple versions of his lecture notes during the 1720s and 1730s. Musschenbroek (1734) is his *Elementa Physicae*, which appeared in several editions. Musschenbroek (1739) is an expanded French translation, published as *Essai de Physique* and reviewed in the *Journal des Sçavans* in June of that year. Musschenbroek (1744) is an expanded English translation. The *Essai de Physique* (1739) contains a final chapter on fire, not found in Musschenbroek (1744). See also his *Institutiones physicae* (1748).

2 The first two of Newton's rules of reasoning appeared as "hypotheses" in the first edition of the *Principia* (in 1687), and it is not until the second edition (in 1713) that Newton separates out his methodological claims and states them as rules of philosophizing, adding a third rule to the initial two. Newton worked on the rules during the intervening period, including the 1690s (see Cohen, 1978) prior to Keill's Oxford lectures. The third edition of the *Principia* (from 1726) contains an additional fourth rule. Correspondingly, 's Gravesande's Chapter 1 ends by stating Newton's first three rules of philosophizing, while Musschenbroek's Chapter 1 ends with them too, and has a discussion of each; both were published before the third edition of the *Principia*. Pemberton's book came out shortly after the third edition and his Introduction concludes with all four of Newton's rules. Ducheyne (2014b) suggests that the omission of the fourth rule from later editions of 's Gravesande's work is significant. For details of 's Gravesande and Musschenbroek on method, see Ducheyne (2014b) and (2015) respectively, and references therein. For Newton's *Principia* in English translation, see Newton (1999).

3 We know about the presence of Newton's rules in the original version of the *Foundations* because, although this version of Chapter 1 does not survive, later chapters of the original manuscript do survive and there we find a cross-reference to a discussion of Newton's third rule (see Barber, 2006, p. 21). See also Janik (1982, p. 99) for a discussion of the drafts of Chapter 1, and more generally for a comparison of the manuscript materials with the published version of the *Foundations*. Barber (2006, p. 21) writes that the original version "seems to have begun with a statement of Newtonian principles of scientific enquiry", and to have been "more concerned with scientific method than with any form of metaphysical doctrine", but his view is that the discussion of method was discarded in favor of chapters on metaphysics. My view is that, despite the radical change in content, the purpose of the early chapters remained unchanged.

4 Newton writes in the General Scholium introduced at the end of Book 3 in the second edition of the *Principia* of 1713 (1999, p. 943):

> I have not as yet been able to deduce from the phenomena the reason for these properties of gravity, and I do not feign hypotheses. For whatever is not deduced from the phenomena must be called an hypothesis; and

> hypotheses, whether metaphysical or physical, or based on occult qualities, or mechanical, have no place in experimental philosophy.

For discussion of the widespread negative association in England of hypothesizing with speculative metaphysical system-building, see Anstey (2005).

5 Musschenbroek (1734) contains a curt dismissal of reasoning via hypotheses, and Musschenbroek (1739) contains a long discussion about the perils of "suppositions". In discussing Bacon's idols, Pemberton offers similar remarks (1728, p. 6):

> all men are in some degree prone to a fondness for any notions, which they have once imbibed, whereby they often wrest things to reconcile them to those notions, and neglect the consideration of whatever will not be brought into an agreement with them.

6 Quotations are from the 1740 first edition of Du Châtelet's *Foundations of Physics*, translated from the original French into English. A complete English translation is available via three sources. See Zinsser and Bour (2009) for the Preface and Chapters 1, 2, 4, 6, 7 (partial), 11 (partial), and 21 (partial). See Patton (2014) for Chapter 9. For the remaining chapters and passages, as translated by Brading *et al.*, see Du Châtelet (2018).

7 For example, Keill states the role and importance of method for this goal clearly and succinctly in his opening chapter (1720, p. 7): "But that we may proceed in this Affair with the greater safety, and, as much as possible, avoid all Errors: we shall endeavour to observe the following Rules".

8 One hazard of translation is that in French the same word "expérience" is used for both experience and experiment, and the reader should be aware of this in reading the quotations from Du Châtelet that follow.

9 Hagengruber (2012b, p. 3) writes that Du Châtelet's "methodological contribution is to establish a new metaphysics, which satisfies the demands of rationality as well as the standards of the experientially dependent contents". I read Du Châtelet's two-pronged methodology as being a contribution to scientific methodology, primarily directed at physics. I agree with Detlefsen (forthcoming) that Du Châtelet is concerned with hypotheses aimed at truth, rather than hypotheses in the sense of devices used for calculation and prediction, as found in positional astronomy.

10 This terminology is anachronistic, but the meaning it conveys to 21st-century philosophers usefully indicates the subject of Du Châtelet's discussion.

11 As is obvious, I am skipping over Chapter 2 (on God) and Chapter 3 (on essences, attributes, and modes). I believe that these chapters are productively read as contributions to the methodology that she is presenting in Chapters 1–4. In brief, Chapter 2 makes clear when and where it is legitimate (and illegitimate) to appeal to God in explanations, and Chapter 3 offers an account of essences, attributes, and modes designed to meet the dual demand of being consistent with the requirements placed on scientific explanations in Chapters 1 and 2, and allowing for causal action between things in the world. I will not make that case here, but for some brief remarks see Section 2.2, as well as *Chapter 4*. For some discussion relevant to these issues, see Iltis (1977, pp. 33–4). Judith Zinsser, in her talk at the

2016 Boston Colloquium on Du Châtelet, argued that method is a unifying theme not only through the early chapters of the *Foundations* but also more widely in Du Châtelet's philosophical work.

12 Winter (2012, pp. 192ff.) made the important discovery of what she believes to be an "advance copy" of the 1740 edition of the *Foundations* sent to Frederick of Prussia. Crucially, the final paragraph of the Preface is a *handwritten* addition to this version, which moreover lacks the footnote referring to Wolff's *Ontology* (subsequently removed in the second edition). It is in this paragraph that Du Châtelet first mentions Leibniz, Wolff, and PSR.

13 Du Châtelet appears to draw on Wolff as well as Leibniz, and Detlefsen (2014) offers a comparison of Leibniz, Wolff, and Du Châtelet on PC and PSR. For Du Châtelet's critique of Locke in relation to her acceptance of PC, see Hagengruber (2012b, p. 9). Du Châtelet's exposition of PSR follows that of Leibniz in *The Leibniz-Clarke Correspondence* (1998), where he interprets Archimedes' account of equilibrium as an example of reasoning using PSR. See also Wolff (2009) Paragraph 10 for his statement of PC and Paragraph 30 for his statement of PSR, as presented in his "Rational Thoughts" of 1720. For discussion of Du Châtelet and PSR, see especially Detlefsen (2014) Section 2, Detlefsen (forthcoming), Moriarty (2006), and references therein. Detlefsen (2014) includes a discussion of the place of PC and PSR in Du Châtelet's methodology, and Detlefsen (forthcoming) characterizes PC and PSR as "rules of reasoning" for Du Châtelet. This characterization seems to me exactly right and highly appropriate: this label highlights Du Châtelet's replacement of a chapter that included Newton's rules of reasoning with her new discussion of PC and PSR.

14 Du Châtelet explicitly rejects perceived clarity and distinctness as criteria for distinguishing between the possible and the impossible, on the grounds that we can be misled about whether or not our ideas have this characteristic. It is only through demonstration of contradiction that we can be assured that we have achieved the necessary clarity for showing that something is impossible.

15 See Biener (2018) and references therein, including Harper (2011) and Smith (2002).

16 Keill (1702, Lecture VIII) explicitly offers constant conjunction and covariation as evidence of a causal connection, and as therefore supporting the case for Newtonian universal gravitation as causal, because satisfying constant conjunction despite involving action-at-a-distance. It is this claim that constant conjunction and covariation provide evidence for a causal connection that Hume attacks in his *Treatise* (1739). Hume wrote the *Treatise* while living in La Flèche, France, from 1734 to 1737. Du Châtelet was becoming more deeply engaged in her own philosophical and scientific work during these years.

17 For more on Du Châtelet's rejection of skepticism as a serious epistemological position, see Janik (1982, p. 105).

18 See, for example, Newton's *Principia*, Book 3, General Scholium (Newton, 1999, p. 940).

19 See also Harth (1992, p. 196). Detlefsen (2014, p. 18), in comparing Du Châtelet to Descartes, attributes a form of nativism to Du Châtelet, such that PC and PSR are innate principles. I read Du Châtelet as arguing for

PC and PSR on the grounds that, without them, knowledge would not be possible, coupled with the presumption that at least some knowledge is, indeed, possible.

20 Janik (1982, p. 104) interprets Du Châtelet as viewing PSR "not as a causal one but as one which precedes and supplements any causal account of physical phenomena". I see it as playing multiple roles for her, including as a constraint on causal explanations.

21 For a discussion of this section of the *Foundations*, including its relationship to the philosophy of Leibniz, see Hecht (2012, pp. 67–9).

22 See *Chapter 3*, Section 3.1, for a brief discussion of the role of PSR and continuity in Du Châtelet's conception of the world as causally interlinked, and in her argument for a strong version of determinism.

23 For discussion of Du Châtelet's views on the existence and nature of God as presented in Chapter 2 of the *Foundations*, see Lascano (2011) and Detlefsen (2014, Section 3). Lascano's discussion shows the importance of Locke as a source for Du Châtelet, and further supports the view that Du Châtelet drew on Leibnizian and Wolffian resources in order to develop her existing project. Terrall (1995. p. 301) remarks that whereas "Leibniz had grounded his philosophy on God's infinite wisdom, Du Châtelet attributes the divine choice to the pleasure God takes in his creation, a pleasure linked to the rational process of optimization... This novel version of the principle of sufficient reason shows God taking pleasure in the rationally determined perfections of his creation", which Terrall suggests is related to Du Châtelet's treatment of happiness in her *Discours sur le bonheur*. According to Janik (1982, p. 111, fn. 11), Du Châtelet dismissed Newton's conception of God's presence in the world with the word "Ridiculous" in the margin next to his claims about the sensorium of God in the General Scholium of the *Principia*.

24 See letter to Maupertuis, 29 September 1738, cited in *Chapter 3*. See also her letter to Frederick the Great of 27 February 1739, where she is worrying about the explanation for the elasticity of bodies. She mentions experimental evidence that she takes to show that subtle matter cannot be its cause (contra the Cartesians), and writes that although Keill appeals to attraction to explain elasticity, she is not convinced that his explanation is satisfactory. Her conviction that appeals to attraction have been taken too far in recent work goes on to be discussed in detail in Chapter 16. She writes that she is very much afraid that we will have to resort to God in explaining elasticity, and that for the time being she will leave physics for geometry, for geometry is the key to all doors and she plans to work to acquire it (doi:10.13051/ee:doc/voltfrVF0900256a1c). Appeals to the will of God as an explanatory resource that appear in the manuscript version of the *Foundations* are removed by the time of the published version (see Janik, 1982, pp. 101–2).

25 For some discussion of the relationship of essences, attributes, and modes to PC and PSR, see Iltis (1977, p. 34). Du Châtelet (2018, 3.47 and 3.50) rejects Locke's view that thought might be added to the essence of matter merely by the will of God (for more on her views of mind and body, see Harth, 1992, p. 197). Similarly, having earlier accepted Newtonian attractive and repulsive forces as qualities of matter (see the first version of her "Dissertation on the Nature and Propagation of Fire"), in the published

version of the *Foundations* she rejects the idea of gravitational attraction as essential to matter (see *Chapter 4*). For more on "thinking matter", and Du Châtelet's importance for La Mettrie, see Hagengruber (2012b, p. 51).

26 See Du Châtelet (2009, 3.49): "One must therefore say that the actuality of things depends on the will of God, for having given existence to this world rather than to another possible world, the World exists, because God wanted it that way". A theme in *The Leibniz-Clarke Correspondence* relating to the voluntarist versus rationalist conceptions of God concerns the manifestation of God's power and wisdom in the created world. Once again, Du Châtelet is now siding with Leibniz. Janik (1982, p. 91) identifies Du Châtelet's rejection of voluntarism and her insistence on rational explanation as an important theme in her philosophy, writing:

> To take refuge in the inscrutable will of God or the limitations of human understanding did not seem to her an adequate response to questions like 'What are the basic constituents of the universe?', 'Why does the law of gravitation hold?', or 'How is human freedom possible in a mechanistic universe?'

On Du Châtelet's rejection of voluntarism, see also Detlefsen (2014, pp. 16–17). For Du Châtelet's engagement with *The Leibniz-Clarke Correspondence*, see Hutton (2012). For further discussion of appeals to the will of God in theorizing, see Hagengruber (2012b, pp. 12, 15, 29).

27 Stan (2018, p. 480) argues that Wolff is the source of Du Châtelet's approach to essence, attributes, and modes. For Du Châtelet's critique of Locke in her discussion of essence, attributes, and modes, see Hagengruber (2012b, pp. 11–12).

28 For an overview of Chapter 1 and its relationship to Chapter 2 (and Du Châtelet's conception of God), as well as to Du Châtelet's discussion of monads, see Erhman (1986, pp. 49–55).

29 For Du Châtelet's presence in the *Encyclopedia*, see Carboncini (1987), Maglo (2008), and Roe (2018). Detlefsen (2014, p. 32) places Du Châtelet "at the forefront of the emergence and embrace of hypothetico-deductive [reasoning] in scientific explanations".

30 For discussions of Du Châtelet on hypotheses more generally, see Hagengruber (2012b, pp. 16ff.), Hutton (2012, pp. 86–7), and Detlefsen (2014 and forthcoming). Detlefsen (forthcoming) provides a comparison of Du Châtelet's use of hypotheses with that of Descartes. I have also benefited from Rey's unpublished manuscript "Une épistémologie originale. Le statut des hypothèses dans les *Institutions de Physique* d'Emilie du Châtelet", material from which will be found in Chapter IV of Rey (forthcoming).

31 Note once again that Du Châtelet's concern is with finding the *true causes* of phenomena.

32 In this, Du Châtelet agrees with Wolff, who writes (1963, p. 67), "Philosophy must use hypotheses insofar as they pave the way to the discovery of certain truth". For Wolff, a hypothesis is an assumption that is made because it provides a reason for certain phenomena, even though its truth has not been demonstrated.

33 Detlefsen (forthcoming) illuminates Du Châtelet's account by identifying similarities and differences between Descartes and Du Châtelet on the

role of hypotheses. There is one additional difference between Descartes and Du Châtelet to which I would like to draw attention. Detlefsen (forthcoming) sees Du Châtelet's starting point as similar to Descartes's in the following respect:

> Like Descartes, Du Châtelet believes that hypotheses are necessary because not all phenomena can be explained through reliance on first principles alone -- there is a gap between first principles and observed phenomena in the world in the sense that the scientist cannot deduce the cause of those phenomena directly or through chains of deduction from the first principles.

I read Du Châtelet differently, as *starting* from the empirical approach of the Newtonians, but finding the methods offered by the Newtonians inadequate for resolving disputes among theoretical alternatives (such as action-at-a-distance versus vortex gravitation and the *vis viva* dispute; see *Chapter 4*). She develops her account of how to use empirical resources to constrain hypothesizing in this context, and only later adds PC and PSR to her overall philosophy (revising her chapter on hypotheses accordingly).

34 In fact, Du Châtelet considers under one umbrella a number of different kinds of hypothesis, including quantitative hypotheses in positional astronomy (such as those due to Ptolemy, Copernicus, and Kepler), the use of hypothetical reasoning in mathematics, and Cartesian speculative hypotheses. See Hagengruber's (2012b) interpretation of Du Châtelet on hypotheses, which depends on this diversity of types of hypothesis.

35 See Ducheyne (2014b) for a discussion of 's Gravesande's attempts to develop a methodology for empirical investigation of hypotheses in the search for certainty. It seems to me that both 's Gravesande and Du Châtelet were troubled by the weakness of the methodological considerations offered by those who rejected Cartesian hypothesizing in favor of a "Newtonian" approach, and that 's Gravesande's discussion is a likely source for Du Châtelet's treatment of hypotheses, which became one prong of her proposed method. Further research is required.

36 I have mentioned the central elements of Du Châtelet's account of hypotheses. More detailed discussions can be found in the references given throughout this section. Du Châtelet's account has some similarities with the brief remarks made by Wolff in Paragraphs 126 and 127 of his *Preliminary Discourse* of 1728 (Wolff, 1963, pp. 67–8). Wolff advocates the use of hypotheses in paving the way to the discovery of truth and emphasizes the importance of experience and experiment in evaluating hypotheses, including their rejection and emendation in light of observations. Du Châtelet's account is not only more developed, but the role and epistemic status of hypotheses within Du Châtelet's philosophy is very different from that within Wolff's, as can be seen by comparison of her account with the extended treatment of hypotheses offered by Wolff in his "De Hypothesibus philosophicis" (1729). See also Vanzo (2015) for a discussion of the role of experience and experiment in Wolff's philosophy.

3 Matter, Body, Force

The subject matter of physics in the early 18th century was bodies: their nature and properties, their behaviors, and the causes and effects of those behaviors. In her *Foundations of Physics*, Émilie Du Châtelet sought an account of bodies as true causal agents in the world. The empirical methods for investigating bodies that she found in the Newtonian textbooks of Keill (1720), 's Gravesande (1720), Pemberton (1728), and Musschenbroek (1734) seemed to Du Châtelet inadequate for this purpose. She articulated a sophisticated method for using empirical resources in theorizing about bodies in what became Chapter 4 of the *Foundations* (see *Chapter 2*, Section 2.3), but even this seemed to her insufficient. Shortly before the publication of the *Foundations*, Du Châtelet supplemented the empirical prong of her method with a second prong drawing on "principles of our knowledge", especially the principle of contradiction (PC) and the principle of sufficient reason (PSR) (again, see *Chapter 2*). This enabled her to develop an account of bodies as causal agents, acting on one another and bringing about change in one another. The details of this account, and the arguments by which she develops it, are our focus in this chapter.

The account that she develops comes at a price, as we shall see. Du Châtelet introduces a realm of the metaphysically real that consists not of bodies but of non-extended simples; bodies become ideal, in the sense that they arise through our experience of that underlying metaphysical reality (about which more later).

At first sight, the account seems metaphysically extravagant compared to the down-to-earth treatments of bodies found in the Newtonian textbooks that were her model. Yet the reasons why she ended up with this account lie in the shortcomings of those textbooks, with respect to the goal of providing an account of bodies as true causal agents in the world. We will see later some of the compelling reasons Du Châtelet had for taking this to be an important goal. The challenge for anyone accepting that goal, while resisting the metaphysical picture that Du Châtelet presents us with, is to find an alternative. I submit that they will be hard pressed to do so. Du Châtelet identified the weaknesses of the Newtonian approach and drew on the best philosophical resources available at the time in order to develop an account that addressed those weaknesses. We will see in *Chapter 4* that she then applied that account to the most difficult and controversial issues concerning bodily action of the time: collisions, gravitation, and *vis viva*. The extent to which she succeeds (or fails) in tackling these problems provides us with further insight into the depth of the philosophical problems facing early 18th-century natural philosophers concerned with bodies as causal agents.

We have seen in the preceding chapters that Du Châtelet's *Foundations of Physics* follows the model of early 18th-century Newtonian textbooks on physics: she begins with method (see *Chapter 2*), and then turns her attention to space, time, matter, bodies, motion, and gravity. In this chapter, we discuss Chapter 7 of the *Foundations*, where Du Châtelet begins her treatment of matter and bodies by addressing the extension of bodies, and then Chapter 8, where she continues her treatment by addressing the forces of bodies.[1]

3.1 The Extension of Bodies

Chapter 7 of the *Foundations* is entitled "The Elements of Matter". The opening paragraphs offer a lightning introduction to the topic through ancient elements (including earth, air, fire, and water), Descartes's matter theory, and atomism. At that time, atomism was the view that matter ultimately consists of "solid, indivisible particles … distinguished from one another by their shape and size" (Du Châtelet, 2009, 7.118[2]). According to Du Châtelet, this account of matter was adopted by most philosophers of the day. Nevertheless, Du Châtelet rejects atomism, arguing instead for non-extended simples as the elements from which a theory of extended matter is to be constructed.

Du Châtelet's rejection of atomism begins as follows (2009, 7.119):

> M. Leibniz, who never lost sight of the principle of sufficient reason, found that these atoms did not explain extension in matter, and, seeking to discover the reason, he believed that it could only lie in a different idea of particles, those without extension, which he named *monads*.

What is the problem with the view that matter consists of atoms of different shapes and sizes, and why introduce non-extended "monads" (whatever they may be) to solve it?

Du Châtelet explicates the reasoning in (7.120–1). The argument is difficult to follow and, in comparison to much of her writing in the *Foundations*, seems rushed and lacking in clarity. However, there are some key points that we can extract from the passages, and we can try to construct a charitable reading of her argument.

The starting point is the claim that bodies are extended: they have length, breadth, and depth. This is common across all accounts of bodies at the time[3] and is supported by experience (all the bodies of our experience are extended). Insofar as any satisfactory physics must recover the bodies of our experience, it must be able to give an account of extended bodies.

One approach might be to say that the extended bodies of our experience are built from atoms: at the time Du Châtelet was writing, atoms were assumed to be small particles of different shapes and sizes, so even though they were very small, they were nevertheless extended; glue a lot of these atoms together and what you get is an extended body. Du Châtelet claims that this appeal to atoms does not work because what we wanted from our explanation was an account of how extended things are possible *at all*. Using atoms fails to meet this explanatory requirement because they are themselves extended: "it is as if one said: *there is extension, because there is extension*" (2009, 7.120). According to Du Châtelet's method in the *Foundations*, more is required: we must show that the idea of an extended body (even a very small extended body, such as an atom) is free from contradiction. And here the atomist runs into a problem.

The atomist view is that atoms are both extended and indivisible. However, it was widely held (not least by Descartes) that whatever is extended is divisible. The atomist must therefore contend with the following three propositions, the conjunction of which yields a contradiction:

1 Atoms are extended.
2 Atoms are indivisible.
3 That which is extended is divisible.

By demanding that we show how extended bodies are possible, Du Châtelet is demanding an account of how it is possible that something that is extended is nevertheless indivisible, without this involving a violation of PC.[4]

One response might be to reject 3, and to insist on the existence of atoms as primitive: while 3 may be a conceptual truth, it is not a metaphysical truth; God made the universe from tiny, extended particles, and he made it the case that they are indivisible, end of story. Indeed, just two years prior to publication of the *Foundations*, Du Châtelet could see no other way forward than this. Writing to Maupertuis on 29 September 1738, she suggested that

> the first particles of matter can be indivisible... by the will of God, for we are often obligated to resort to him, and I believe this indivisibility of the first bodies of matter to be an indispensable necessity in physics.[5]

However, by the time of the *Foundations*, she thinks that we can, and must, do better. In setting out her method, Du Châtelet argues that such appeals to God are not to be admitted in physics (see *Chapter 2*). While God is the source of the actuality of all things (he chose this world, among all possible worlds, and brought it into being), he is not the source of their possibility. In seeking to show how extended bodies are *possible*, appeal to God will not do (2009, 7.121). In Chapter 1, in her discussion of PC, Du Châtelet had warned: "One should be just as cautious when maintaining that a thing is possible; for one must be in a position to show that the idea is free of contradiction" (2009, 1.6, and see *Chapter 2*). But this is precisely what the atomist cannot do, it seems.[6]

If we are unable to show how something that is extended yet indivisible is possible, then the atomist takes a significant epistemic risk: she incorporates into physics, at its very basis, entities that are at best inexplicable and unintelligible, and at worst self-contradictory. Appeal to God, or to brute fact, as the explanatory terminus of the possibility of extended atoms, accepts this risk, but the adoption of PSR as a methodological requirement rules out the taking of such risks, especially when an alternative is available.

Du Châtelet believed that the construction of extended bodies from non-extended simples provides a viable alternative to atomism. The idea of a non-extended simple does not contain within it the risk of self-contradiction so evident in the idea of atoms, and so does not demand further explanation in the way that the appeal to atoms does. This is one reason for preferring non-extended simples over extended atoms as the starting point of physics.[7]

However, Du Châtelet concludes not just that non-extended simples are to be preferred over atoms as the starting point of physics, but that they are to be admitted as *necessary* (7.122). Detlefsen's (2014, pp. 21ff.) helpful analysis of Du Châtelet's argument for non-extended simples is especially useful on this point. First, we assert that extended bodies are composites, and ask about the nature of the elements of which they are composed.[8] Second, from PC, we conclude that nothing that is extended can serve as the elements out of which extended bodies are composed. Therefore, the elements must necessarily be non-extended beings. This conclusion gives us two results: it establishes the *nature* of the elements as necessarily non-extended, and it proves the necessity of non-extended beings as the elements of composite extended beings. Given the premise that composite extended beings (i.e. bodies) exist, the necessary existence of non-extended beings follows. This is a kind of "hypothetical necessity", as Detlefsen explains.[9]

This conclusion satisfies Du Châtelet's requirement that we "arrive at necessary things when explaining the origin of beings", where a "necessary thing" is one for which there is a contradiction "in what is opposed to it" (2009, 7.121). For atoms, as extended things that are nevertheless indivisible, to be necessary, there would have to be a contradiction between being extended and being divisible. But since there is no such contradiction, atoms are not necessary beings. On the other hand, there is a contradiction in extended yet indivisible atoms, and these are what is opposed to non-extended simples; therefore, non-extended simples are necessary things. Once again, however, we need to be clear that insofar as this is an existence claim, the necessity is hypothetical, as Detlefsen stresses. We need to distinguish between the claim that, since there are extended things, there must be simple things, and the claim that such simple things are in themselves necessary. Du Châtelet offers an argument for the former claim in (7.120), but no argument (that I have found) is offered for the latter claim, and so her claims about the necessity of non-extended simples should be interpreted accordingly.

So far, we have seen Du Châtelet's reasons for rejecting atomism and adopting non-extended simples as the basis of her matter theory. The next step is to show how it is possible for extended bodies to arise from non-extended simples. As noted, we are seeking an account of the extended bodies that are the bodies of our experience.

The heart of Du Châtelet's argument for how the extended bodies of our experience arise from non-extended simples is as follows (2009, 7.133):

> Furthermore, it is by this dissimilarity that one can understand how non-extended Beings can form extended Beings; for the Elements exist each of them necessarily external to the others

> (since one can never be the other), and all of them being, as we have just seen, united and linked together, an assembly of several diverse Beings results from this, each of which exists external to the others, and which by their interconnections make a whole; but I have shown that we cannot represent extension other than as an assembly of several diverse, coexisting things, and which exist external to one another (§77): therefore, conclude the Leibnizians, an aggregate of simple Beings must be extended. Thus, from the Metaphysical union of the Elements flows the Mechanical union of the Bodies that we see; for all Mechanics that falls under our senses derives in the end, and in going back to the first source, from the superior and Metaphysical principles.

Although Du Châtelet refers to "the Leibnizians" here, her position differs in interesting ways from those of Leibniz and Wolff, who were her "Leibnizian" sources.

We can reconstruct her argument as consisting of two premises and a conclusion.

Argument from simple beings to extended bodies

(P1) Bodies are composite beings arising from multiple coexisting simple beings.

(P2) We necessarily represent a multiplicity of coexisting things as spatially extended.

(C) We necessarily represent (i.e. represent to ourselves) composite beings (i.e. bodies) as spatially extended.

This was not quite the conclusion we had in mind, for it makes the extension of bodies a function of our perceptive faculties. To better understand the argument, and the conclusion to which it leads, we need to unpack each element of the argument in turn. Premise (P1) contains Du Châtelet's views on the interconnectedness of simple beings, which are intimately related to her views on the law of continuity and determinism; premise (P2) depends on her views on space; and the conclusion (C) is to be understood in light of her position on our perceptive faculties in relation to simple beings.

3.1.1 The Interconnectedness of Coexisting Simple Beings

We arrive at premise (P1) of Du Châtelet's "Argument from simple beings to extended bodies" from statements that she makes in Chapter 7.

(P1) contains the claims that bodies are *composite*, that they *arise from* simple beings, that they arise from a *multiplicity* of simple beings, and that these simple beings from which they arise *coexist*.

In (2009, 7.120), Du Châtelet states first that bodies are extended beings, and then that "a being without extension and without particles is a simple being; so, compounds, extended beings, exist because there are simple beings". In (7.125), Du Châtelet asserts that since simple beings are the origin of compound beings, the sufficient reason for all that is found in compound beings must be found in simple beings. Later, she also says (7.131), "The original reasons for all that happens in bodies lie necessarily in the elements of which they are composed". These remarks confirm that bodies, as extended things, are *composite*, and that bodies depend in some way on (or *arise from*) non-composite, simple beings, or elements. They also implicitly confirm that the simples out of which bodies arise coexist, and I have made this coexistence explicit in (P1).

For Du Châtelet, coexistence implies interconnectedness (7.130): "All is linked in the world; each being has a relationship to all the beings that coexist with it".[10] In fact, this interconnectedness is extremely strong (see 7.130–1):

> For the state of any element A being determined, harmony and order require that the state of its neighbors B, C, D, etc. should also be determined in a particular manner rather than in any other, to work in harmony with the state of the first; and as the same reason continues for all states of all the elements, all future states of the elements will also have a relation to the present state that must coexist with them, to past states from which this present state results, and to the states that will follow it, and of which it is the cause. Thus, it can be said that in M. Leibniz's system, it is a metaphysical-geometrical problem, the state of an element being given, to determine the past state, present, and future of all the universe.

This is a remarkable statement of a very strong form of determinism, predating Laplace's famous 1814 statement of determinism by 64 years. Du Châtelet's position follows, she argues, from PSR and the law of continuity. Much though it is tempting to digress into a discussion of PSR, continuity, and determinism, and Du Châtelet's role in this important issue in philosophy of physics, we will stick to our narrow path, noting only that the relevant articles in the highly influential *Encyclopedia* of Diderot and d'Alembert (on continuity and so forth) were taken from Du Châtelet's *Foundations of Physics.*[11] For our purposes, the important point is that the simples, out of which bodies are "composed", are highly interconnected with one another: they stand in deterministic causal relations (though not, as we shall see, in spatial relations) to one another.

Du Châtelet says that simples "compose" bodies (7.131). However, whatever this "composition" relation is, it is not straightforwardly one of mereological parts and wholes (it is not that by taking more than one simple we get a body), nor is it the case that gluing the simples together in causal relations is sufficient to compose a body. Coexistence and causal interrelatedness of the simples (as set out in premise (P1)) are necessary conditions for a body to "arise" from the multiplicity of simples, but they are not sufficient: also in the mix is the role of our perceptual faculties, and this is premise (P2).

3.1.2 Space and Body

Premise (P2) of Du Châtelet's "Argument from simple beings to extended bodies" asserts that we necessarily represent a multiplicity of coexisting things as spatially extended. In order to understand this premise, we must turn our attention back to Chapter 5 of the *Foundations*, and to space.

Du Châtelet's chapter on space begins by discussing whether space is absolute or relative. Du Châtelet briefly rehearses arguments from *The Leibniz-Clarke Correspondence* (see Leibniz and Clarke, 1998) and sides with Leibniz's endorsement of a relational view of space on the grounds that absolute space is ruled out by PSR (2018, 5.74). We will not review these arguments here, for there is a large literature on the absolute versus relative space, time, and motion debate in philosophy of physics, and Du Châtelet's position on these issues can be assessed largely independently of our purposes here.[12] However, there is one particular passage in the discussion of space that is crucial for her account of bodies, for it supplies (P2) in the previous argument. Suppose that we wish to imagine *two* similar apples. How do we ensure that we have *one* representation of *two* apples, rather than *two* representations of *one* apple? The answer, Du Châtelet suggests, is that we represent the apples to ourselves as standing in spatial relations to one another. The spatial relations ensure that we have a single representation (a unity) of a two things (of a diversity or multiplicity). The origin of our idea of space, Du Châtelet argues, lies in our representation of multiplicities. She writes (2018, 5.77), "we cannot represent to ourselves several different things as being one, without this resulting in a notion that is attached to this diversity and union, and this notion we call *Extension*". In order to represent a multiplicity of things, we necessarily represent them as standing in spatial relations to one another, with the upshot that the multiplicity, taken as a whole, is spatially extended.

This, then, is the second step in the argument for how it is possible that bodies are extended. We begin with a multiplicity of causally ordered, *but not spatially ordered*, non-extended simples. Bodies, as extended objects, arise from multiplicities of non-extended simples through how we represent those multiplicities, viz. *spatially*.[13]

This account involves an interesting strand of idealism about both bodies and space.[14] Bodies are ideal in the sense that they are spatially extended entities arising through our representation of underlying simples that are themselves nonspatial: simples are neither extended nor spatially ordered in any way. Space itself is similarly ideal. As mentioned, in Chapter 5 Du Châtelet discusses *The Leibniz-Clarke Correspondence* (1998) and sides with Leibniz in rejecting absolute space. However, while Du Châtelet endorses the Leibnizian view that "space is the order of co-existing things" (2018, 5.79), for her this is not a straightforward relationalism about space. Rather, the order that is being represented spatially is the *causal* order of the coexisting things, not a spatial ordering: the simples are not themselves spatially ordered; the spatial representation is *our* representation of that causal order. Space is, in this sense, ideal. We can approach the same point another way: space is relational in the sense that it arises from the relations among bodies, but since the spatiality of bodies arises from how we represent a nonspatial underlying reality, space is not only relational but also ideal. There is a contrast with the case of time that is helpful. Du Châtelet follows Leibniz in rejecting absolute time (see Chapter 6) and in endorsing the view that "time is really nothing other than the order of successive beings" (2009, 6.102). She asserts an analogy between space and time (2009, 6.94), but I think we can also see an important *dis*analogy: the successive states of the simples stand in temporal relations to one another (see 2009, 7.128–31), arising from the causal order independently of our experience of them, whereas spatial relations arise only through how we experience and represent that causal order. Time may be relational for Du Châtelet, but it is not ideal in the way that space and bodies seem to be.[15]

3.1.3 Confused Perceptions

The conclusion of Du Châtelet's "Argument from simple beings to extended bodies" asserts that we necessarily represent (i.e. represent to ourselves) composite beings (i.e. bodies) as spatially extended. One might object that (P1) and (P2) nevertheless do not yield the bodies of our experience: bodies do not look to us as though they are multiplicities of non-extended simples, so how can it be correct to say that it is through our representation of these multiplicities to ourselves, i.e. through our perception of them, that we see bodies as extended? Paragraph (7.134) addresses this issue. The answer is that we are unable to perceive the multiplicity distinctly, as a collection of interconnected simples, perceiving them only confusedly instead. Confused perception accounts for how a collection of interconnected simples yields the experience of a single extended body with various qualities pertaining to the whole.

Du Châtelet offers an extended treatment of this issue, but a short quotation here will have to suffice (2009, 7.134):

> But as it is impossible that we represent to ourselves the internal state of all the simple Beings (upon which, however, the phenomenon of extension depends), all perception of realities must escape us by our nature; and there remains to us nothing but the confused ideas that we have of each of these simple Beings, which is an idea of several coexisting things, linked together without us knowing distinctly how they are linked, and it is this confused idea that brings into being the Phenomenon of extension.

If we could perceive each of the simples individually, and if we could distinctly perceive their relations to one another, the phenomena of extended bodies would disappear. But we cannot. The upshot of our confused perception is a blurring over, yielding the phenomena of extended bodies and their qualities.

We can see how Du Châtelet's position differs from that of Leibniz and of Wolff. For Leibniz, monads (his simples) do not causally interact with one another (they are "windowless"), but they are ordered and each represents that order (with more or less clarity) according to preestablished harmony. For Wolff, the simples are ordered, and he seems indecisive about whether simples act upon one another.[16] Du Châtelet commits herself, as we have seen, to causal interaction among the simples, and to their causal ordering. Watkins (2006, pp. 283–4) offers the following comparison of Leibniz and Wolff on space and extension. Leibniz explains space as arising within a monad's representations, and from the confusion of those representations, such that only a single monad and its representations are required for the explanation of extension and of extended bodies. Wolff, on the other hand, maintains that space is an ordering relationship among coexisting simples, arising from their intrinsic qualities, so that a plurality of simples is required for the explanation of extension. He writes, "as soon as we represent this order to ourselves, we represent space to ourselves" (see Watkins, 2009, p. 15, for the relevant passage in Wolff's *Rational Thoughts*). Wolff argues that since a plurality of coexisting simples cannot exist at the same point, that plurality therefore yields a composite that is spatially extended, i.e. a body.[17] For Du Châtelet, our experience of extended bodies involves two steps. First, a plurality of causally related simples provides the explanation of extension: that we represent the simples as external to one another is sufficient for extension. However, the representation of two simples (say) as external to one another and standing in relations to one another would not yield the phenomena of bodies, for those simples would remain distinct from one another in the representation. In the second step, the confusion in how we represent the plurality of causally related simples to ourselves yields our experience of that plurality as a *single* extended thing (rather than as a plurality of simples) with qualities pertaining to that thing as a whole. This completes Du Châtelet's account of how it is that the extended bodies of our experience arise from non-extended simples.

Before moving on, there is one final puzzle in her account of bodies to address. We have seen that for Du Châtelet, bodies arise through how we represent multiplicities of interconnected, coexisting simples to ourselves. However, for Du Châtelet, all simples are causally interconnected with one another, and this seems to yield a problem for her account. We experience the world as containing a multiplicity of bodies. How, if at all, is the totality of coexisting simples divided into the parts from which arise the multiple bodies of our experience? Du Châtelet provides no principle for the division of her causally interconnected system of simples into subsystems from which bodies might arise.

It seems that the division of the world into a multiplicity of bodies must take place within our experience of the world, through how we represent the simples to ourselves, rather than within the causal relations among the simples themselves. This is consistent with Du Châtelet's view of the world as a plenum (see Chapter 5), and with her account of bodies in Chapter 10. At the beginning of Chapter 5, in her discussion of absolute versus relative space, Du Châtelet rejects the idea that there is any empty space, be it between bodies or within the "pores" of bodies: she accepts the plenum and offers responses to arguments against the plenum that were common amongst antiplenists at the time. In Chapter 10, she distinguishes between fluids, soft bodies, and hard bodies by appealing to the relative motions of the parts of those bodies, the pores of a body plus the passage of fine matter through those pores, and the ease with which the parts of bodies yield to our efforts to separate them.[18] These distinctions seem to have more to do with how *we* experience the material world, than with principled divisions within the plenum.

3.1.4 Back to Atoms

For Du Châtelet, the extended material world is a plenum, divided into bodies through how we experience that plenum. Nevertheless, in 1738 she had written to Maupertuis that the indivisibility of "the first bodies of matter" is "an indispensable necessity in physics" (see Section 3.1). In subsequently rejecting atomism, it may seem as though she has embraced the divisibility of matter all the way down, and is no longer entitled to indivisible "first bodies" in her physics. Not so, as it turns out. Having shown how extended bodies are possible (Chapter 7), in Chapter 8 she further develops her discussion of the nature of bodies (see later), and then in Chapter 9 she turns to the divisibility of matter. She proceeds to use the account of the extension of bodies given in Chapter 7 to argue for the indivisibility of physical atoms. If this argument is successful, then Du Châtelet can have her cake and eat it: she can avoid the difficulties with primitive atomism (see Section 3.1) while availing herself of its explanatory power.

At the heart of her argument is a distinction that she makes between geometrical bodies and physical bodies. Geometrical bodies are the mathematical objects that we consider when doing geometry; they are the abstractions of our mind (Du Châtelet, 2014, 9.166), and we can divide them at will. Geometrical bodies, she says, contain no actual parts but merely potential parts[19] (9.169), and are thus divisible to infinity: "The geometrical body has no determinate and actual parts, it contains nothing but simply possible parts, which one can increase to infinity as one wills". The situation with physical bodies is very different. Physical bodies have actual parts, and we are not free to divide them as we please. This is because every body in nature arises from a finite and determinate number of simples, and all the actual parts of a body must also arise from a finite and determinate number of simples. So, the only divisions of the body into parts are those that yield a determinate and integer number of simples for each part. Therefore, bodies are only finitely divisible (9.170). Moreover, a multiplicity of simples is required for extension, and therefore every actual part of a body arises from a multiplicity of simples. Therefore, there are least parts of matter: they arise from the simples but cannot be further divided into those simples without ceasing to be extended bodies. Moreover, Du Châtelet suggests that nature may be such as to never further divide its smallest bodies into those simples, so that there are stable and determinate smallest parts of matter: atoms. She concludes (9.171): "Thus, the divisibility of extension to infinity is at the same time a geometrical truth and a physical error". In sum, Du Châtelet rejects atoms as primitive, develops an account of the possibility of extended bodies based on non-extended simples, and by means of this account argues for the possibility of stable least parts of matter: derivative physical atoms.

This highly interesting argument rejects the widely held view that the infinite divisibility of geometrical extension shows the possibility (at least) that physical extension is infinitely divisible. Rohault (1671) argued for the infinite divisibility of matter in this way in his Cartesian textbook; Leibniz accepted this view in rejecting Descartes's view of matter as extension; Newton accepted this view in considering the divisibility of body and suggesting that the issue is to be resolved empirically (Brading, 2018b); and the problem of divisibility is discussed in several of the standard Newtonian textbooks.[20] Du Châtelet's argument is also relevant to discussions at the time about the roles of mathematics in natural philosophy, and more specifically the extent to which inferences about the physical can be drawn from features of the mathematics used in theorizing about the physical.

The arguments thus far have been drawn entirely from one prong of her method: the principles of our knowledge. We have seen both PC and PSR at work. However, as we saw in *Chapter 1*, there is a second prong to her method, and she brings this to bear on the question of the divisibility of matter. She writes (2014, 9.171):

> That nature halts in the analysis of matter at a certain fixed and determinate degree, is sufficiently probable, on the basis of the uniformity that rules nature's works, and on an infinity of experience.

She then reviews the empirical evidence for atomism, which had seemed to her to show atoms indispensable in physics.

From one prong of the method, we have shown how atoms are possible, and from the other prong, we have shown that they are probable, and this is as far as we are able to proceed.

The first stage of Du Châtelet's account of bodies is concluded. She has shown how it is possible for bodies to be extended. In the next section, we see what more must be done in order to complete the account of bodies.

3.2 The Force of Bodies

Chapter 8 of the *Foundations* begins with a brief review of the Cartesian position that extension is the essence of body, and then Du Châtelet makes the following remark (2018, 8.138):

> This definition of the essence of a Body drove them necessarily to remove all force and all activity from creatures; for however we reflect upon extension, limiting it in whatever ways we would like, or arranging its parts in every way possible, we do not at all see how there can arise from it a force and an internal principle of action.

This leaves the Cartesians with a problem, Du Châtelet argues, because "experience proves that bodies act and are gifted with activity" (8.138), but their account of body lacks any resources for the provision of this activity, and thus their only recourse is to God, and to occasionalism[21]:

> Thus, according to them, it is not the creatures that act, it is God himself who immediately moves a Body on the occasion of another... Thus, secondary causes do indeed seem to have some efficacy in this system, but in reality they have none. God does everything by his immediate concourse: the creatures are the occasions, but never the causes; they can receive, but they can never act, nor produce.

The question of how it is that bodies are capable of acting on one another is, I think, *the* central and driving question of the *Foundations*. Yet Du Châtelet would have found little help on this issue in the Newtonian textbooks that she studied. Musschenbroek is perhaps the most pessimistic, and also the most explicit (1744, Preface, p. ix):

> we do not understand the manner of operating of any one thing; and all we can do is to observe the effects that constantly flow thence. When two bodies impinge against one another, how do they operate upon each other? What is the force, and how is it transferred from one to the other?... All these mysteries are concealed from us mortals.

Her difficulties with this issue are reflected in the manuscript version of her book, where she acknowledges that she has come up against an impasse, and appeals to God[22]:

> In the end it seems to me that it is no easier to conceive the simple communication of movement between bodies supposed to be

> completely hard, than to know what their forces will be after the collision; one must, I think, leave both questions to God.

She goes on:

> The simplest case of them all is that of one body that hits an immovable obstacle, and this case is subject to the greatest difficulties. I am quite afraid that we must resort to God for the collisions of bodies.

Ultimately, however, Du Châtelet found the appeal to God as the solution to the problem of bodily action unsatisfactory. In particular, this approach carries the risk of yielding bodies without genuine causal agency after all, which was her reason for rejecting the occasionalism of which she accused the Cartesians. Moreover, it leaves the agency of human beings a mystery: without an account of how it is that physical bodies act on one another, we cannot proceed to an account of how we, as embodied agents, act freely in the world. Du Châtelet was deeply involved in a discussion of the appropriate measure for the force of bodies (this is the so-called "*vis viva* controversy", which we discuss in *Chapter 4*), where the "force" of a body was thought of as the means by which it acts on another body. This dispute included a discussion over whether the total amount of force in the universe is conserved, and Du Châtelet worried about the implications of conservation of force for the possibility of human free will. On 30 April 1739, she wrote to Maupertuis on exactly this issue (see Bour and Zinsser, 2009, p. 109):

> But the only thing that puzzles me at present is liberty, for in the end I believe myself free and I do not know if this quantity of force, which is always the same in the universe, does not destroy liberty. Initiating motion, is that not to produce in nature a force that did not exist? Now, if we have not the power to begin motion, we are not free. I beg you enlighten me on this point.

We know that free will was an important issue for Du Châtelet, not just from this concern expressed to Maupertuis, but also because she wrote an essay on the topic. Her essay "On Liberty" was originally intended for inclusion in the *Foundations*,[23] and it argues that two elements are required for free human action: a will governed by reason, and bodily agency. The point about bodily action is crucial. She writes (Voltaire, 1989, pp. 493–4, my translation) that "the will is never the cause of our actions, even if it is the occasion of our actions" and that the "physical power to act is thus what makes man a free being". The question of bodily agency, of how it is that one body acts upon another, is therefore of central importance to Du Châtelet's philosophical concerns. For Du Châtelet, the Newtonian method had failed to deliver on the most pressing problem of all left in the wake of Newton's *Principia*: the problem of human action in the world.[24]

I believe that the new methodology, introduced in the revisions to the *Foundations* immediately prior to publication (see *Chapter 2*), was introduced precisely because it enabled Du Châtelet to address the problem of bodily action, and that this is what drove the revisions to the text.

Having claimed that appeal to the extension of bodies is insufficient for providing an account of how it is that bodies act on one another, Du Châtelet opens a second line of attack: any account which admits only extension to the essence of body violates PSR, via the principle of the identity of indiscernibles (PII), since such matter would be entirely homogeneous and so all parts of matter would be similar (2018, 8.139).[25] On the basis of PSR, we conclude that there must be something more to the essence of matter, such that the parts are discernible: "There must, therefore, be something in Matter from which this internal difference originates" (2018, 8.139).

Du Châtelet moves immediately to assert that by the addition of "force" to the essence of matter, we ensure that PSR is satisfied. The argument for this conclusion is extremely compressed (see the second paragraph of 8.139), but it seems to proceed along something like the following lines.

Suppose that matter were purely extension. Suppose that the parts of a portion of this matter, however small, were all at rest. Then they would be entirely similar. But by PSR, this cannot be the case. Therefore, all the parts of matter, however small, must be in different states. Moreover, and crucially, in order to satisfy PSR the properties that differentiate one part of matter from all the others must be "in" that part of matter: there must be "a real difference between all the parts of Matter". The source of this real difference is an "internal force".

The introduction of a force in all parts of matter seems to be a bit of a leap, but we can interpret it modestly as being whatever is needed in order for a part of matter to be always in its own distinct (though perhaps changing) state. Neatly, this is also going to be what provides a body with the power to act. If we accept that all changes in bodies are rearrangements of parts of matter, then the power of one body to act on others is its power to rearrange them (or their parts) through motion. Du Châtelet identifies the force of a body, introduced in order to satisfy PSR, with the force by which it is able to act on others: the "internal force" is therefore a "force tending toward motion" (8.139), a "motive force" (8.141).[26]

The addition of motive force as an essential property of matter does not complete the account of bodies, and of how they act on one another. Du Châtelet argues, on the basis of both reason and experiment, that matter has also a passive or resisting force associated with it, for how can one body act on a second, unless the second body resists the action of the first (8.142)[27]:

> Reason shows us and experience confirms for us another property of Bodies, that of resistance, or passive force; for in reasoning from the active force that is in Bodies, we do not see what it would act upon if Bodies did not have resistance, since there would then be no sufficient reason for their action.
>
> On the other hand, everyday experience confirms that when we want to set a Body in motion that seems to us to be at rest, we cannot achieve this without an effort... This Body therefore has a force by which it resists any motion that we want to impart to it.

Du Châtelet argues that all changes that happen in bodies can be explained by appeal to extension, active force (or motive force), and passive force (or resisting force), and that these three principles are mutually independent (8.147–9) and jointly necessary and sufficient (8.145) for an account of the nature of body. Such bodies are capable of acting on one another, and of being acted upon.[28] The essence of body therefore consists of three mutually independent and jointly necessary principles: extension, active (or motive) force, and passive (or resisting) force.

The forces of bodies do not, in fact, appear out of nowhere. We know from Chapter 7 that bodies arise from the simples out of which they are, in some sense, composed. Moreover, the properties of bodies depend upon (or supervene on, we might say) the properties of the simples in the following sense (2009, 7.125): "the sufficient reason for all that is found in compound beings must be found in simple beings". Therefore, the changes that we see around us in bodies must have their origins in simple beings:

> Perpetual change can be observed in compounds; nothing stays in the same state; all tends to change in nature. Now, since the primary cause for what happens in compounds must ultimately be found in simple beings, from which the compounds resulted, there must be found in simple beings a principle of action capable of producing these perpetual changes....

From here, she concludes as follows (7.126):

> The principle that contains the sufficient reason for the actuality of any action is called force; for the simple power or ability to act is only a possibility of action or of passion in beings, which requires a sufficient reason for its actuality.

This is her definition of force: the sufficient reason for the actuality of any action. She goes on to further explain the distinction between the power that an animal (say) has to act (e.g. to walk), and the actualization of this power (an animal that has the power to walk need not always be walking). Force is the sufficient reason for the actualization of a power.

The actualization of a power can be prevented by resistance (7.126):

> Now, the sufficient reason for all that happens to compounds lying ultimately in simple beings, it follows that simple beings have this force, which consists in a continual tendency to action, and this tendency always has its effect when there is no sufficient reason preventing it from acting, that is to say, when there is no point of resistance; for one must call resistance that which contains the sufficient reason why an action does not become actual, though the reason for its actuality remains.

According to Du Châtelet's account, "the force of simple beings is deployed continuously" (7.126), as a result of which each simple being undergoes a continuous succession of changes. Simple beings are, in this way, active, with a "force which to them is essential, which never leaves them, and which cannot cease" (7.128). The resulting sequence of change is unique to each simple, and thus force is the principle by which simples satisfy Leibniz's PII. This account consists of a series of assertions with little by way of supporting argument or evidence. However, our purpose here is not to critique this account, but to see how Du Châtelet deploys it in solving the problem of action among bodies. For Du Châtelet, just as the extension of bodies is a phenomenon, arising from how we experience the simples, so too are the active and passive forces of bodies: they are phenomena arising from the active and passive forces of simples, through how we experience those simples (8.155–7). Referencing Leibniz, Du Châtelet terms the active force of simples primitive force, while the active force of bodies is derivative force (8.158).

Both Leibniz and Newton had argued for the reintroduction of some notion of force into our account of matter and body. In his *Principles of Philosophy* of 1644, Descartes had sought to remove any notion of force from physics (see Descartes, 1991), and in his studies on motion Galileo had done the same. At the time Du Châtelet was writing, it was controversial whether any such notion was needed, and unclear what roles and definition any notion of force should have. Force in Newton's *Principia* is a complex and multifaceted concept for which no general definition is given: instead, Newton defines different types of force, including the force of inertia (even though with hindsight we no longer call inertia a force). Du Châtelet found the ideas of force available in the writings of the Newtonians insufficient for her purposes. She drew instead from the Leibnizian resources discussed earlier. She sought to meet the force needs of the Newtonians using these resources and did so by, for example, reframing Newton's laws of motion in these terms (see Du Châtelet, 2009, Chapter 11; Reichenberger, 2018; and *Chapter 4*). Since Newton explicitly eschewed crucial distinctions in the Leibnizian framework (such as that between active and passive force), the task is fraught with difficulties. A crucial question is whether Du Châtelet's account is adequate for the needs of a successful physics. We continue our discussion of Du Châtelet's account of force in *Chapter 4*, where it is put to work in the most important cases of bodily action in physics at the time. Our purpose in this chapter has been to set out the account of bodies on which she will rely.

Du Châtelet's account of bodies satisfies her methodological requirements as follows. By means of experience, Du Châtelet has argued that there are extended bodies that act and are acted upon by one another. By means of PC and PSR, she has argued that there are non-extended simples endowed with an active and a passive force by which they are able to act and be acted upon, and by which, through the active force, they satisfy PSR (via PII). By means of these resources, she has shown how it is possible for there to be the bodies of our experience: extended bodies that are capable of acting on one another.

I have claimed that the new methodology, introduced in the revisions to the *Foundations* immediately prior to publication, was introduced precisely because it enabled Du Châtelet to address the problem of bodily action, and that this is what drove the revisions to the text. In order to address the problem, Du Châtelet sought to answer this question: what is the nature of bodies such that they are capable of acting on one another? In Chapter 7, she showed how it is that extension belongs to the nature of bodies. She then had to show what else, if anything, belongs to the nature of bodies such that they are capable of acting on one another. She did this in Chapter 8, where she introduced active and passive force. Critically, both of these chapters depend on the new methodology, and especially on PC and PSR, in order to develop their solutions.

Moreover, it is the employment of PC and PSR in their methodological role, in attempting to solve the problem of bodily action, that leads to the introduction of the central metaphysical commitments of the *Foundations*, including non-extended simples in her account of extended bodies, and her complex theory of forces, including the primitive force of non-extended simples, in her account of the agency of bodies.

There is no doubt that Du Châtelet has been led to introduce a rich set of metaphysical commitments into her account of bodies. However, they have been introduced in order to provide an account of the bodies that are the subject matter of physics such that these bodies are themselves causally efficacious, acting on and being acted upon by one another. For Du Châtelet, a physics that is unable to provide an account of its own subject matter is dangerously incomplete. While we might be tempted to say, "so much the worse for physics", Du Châtelet recognized the wider implications of such a response. Physics *is* the science of bodies, and insofar as we are embodied agents in the world, any area of philosophy that includes human bodies within its subject matter presupposes that an account of such bodies is to be had. If there is no such account, then these other areas of philosophy also have a dangerous lacuna. All areas of philosophy that incorporate a notion of human bodily agency, including moral and political philosophy, incorporate an undischarged promissory note, and no amount of waving a hand in the direction of some commonsense notion of body will fill the void. Du Châtelet uses PC and PSR as powerful tools for pinpointing the epistemic risks of proceeding otherwise than she suggests, and for showing the kinds of commitments that will be needed in order to avoid those risks. If you find yourself resisting the metaphysical commitments that she introduces, the onus is on you to suggest an alternative and argue for its superiority.

I have argued here that Du Châtelet's metaphysical commitments have been introduced with a specific philosophical purpose in mind. This is important for how we read the *Foundations*, especially in our attempts to recover a text that has yet to be given extended treatment by philosophers. Specifically, the interpretation allows us to assess the strengths and weaknesses of Du Châtelet's arguments with respect to her own philosophical goals.

In short, my claim is as follows:

- the problem of bodily agency is a central and pressing problem for Du Châtelet;
- she finds the Newtonian method to be impotent with respect to this problem;

- shortly before publication, she found a new method that had resources to make progress with the problem;
- the use of this method to address the problem is what led to the introduction of a range of new metaphysical commitments into the philosophy of the *Foundations*.

In this chapter, we have seen how Du Châtelet began to solve the problem of bodily action, that I claim lies at the heart of the *Foundations of Physics*. She did so by showing how it is possible for there to be bodies acting on one another. The next step is to use this account of bodies to address the three most controversial issues of bodily action in physics at the time: collisions, gravitation, and *vis viva*. We turn our attention to these in *Chapter 4*.

Notes

1 See Appendix 1 for a complete list of chapter headings of the *Foundations*. We have skipped over the chapters on space and time, and will not provide an independent discussion of them in this book. I refer back to them in what follows, as needed in discussing Du Châtelet's arguments concerning matter, body, and force. Du Châtelet's chapters on space and time deserve treatment in their own right. For a brief discussion see Suisky (2012, pp. 138–9).

2 Quotations are from the 1740 first edition of Du Châtelet's *Foundations of Physics*, translated from the original French into English. A complete English translation is available via three sources. See Zinsser and Bour (2009) for the Preface and Chapters 1, 2, 4, 6, 7 (partial), 11 (partial), and 21 (partial). See Patton (2014) for Chapter 9. For the remaining chapters and passages, as translated by Brading et al., see Du Châtelet (2018).

3 For the seeds of a concept of body where body may lack extension, see Biener and Smeenk's (2012) discussion of Newton's geometrical versus dynamical conceptions of body in his *Principia*.

4 See also Detlefsen's (2014, p. 21) very helpful account.

5 For the problem of divisibility in 17th- and 18th-century matter theory, see Holden (2004).

6 The atomist might attempt to strengthen her claim by asserting that her atoms are *necessarily* both extended and indivisible. This will not do, Du Châtelet argues, because the claim of necessity is to be demonstrated by showing that its opposite leads to a contradiction (see *Chapter 2*), and there is no contradiction in things that are extended and yet divisible.

7 For Leibniz's argument from the divisibility of extension to the introduction of simples, see Watkins (2006, pp. 267ff.). Garber (2009) has termed the problem "division-to-dust", but, as I have noted elsewhere (Brading, 2018a), this seems too weak to convey the problem that Leibniz sought to solve. Watkins notes that Wolff added a second argument for the introduction of non-extended simples. It seems to me that Du Châtelet adopted a version of Leibniz's argument, to which I have here given a formulation

that, in line with Du Châtelet's emphasis on the methodological roles of PC and PSR, places emphasis on PC and PSR in guiding our rejection of extended atoms and our adoption of non-extended simples as the basic entities out of which bodies are composed.

8 Du Châtelet asserts that bodies are composites, rather than arguing for it. Detlefsen (2014, p. 21) suggests that it follows from the claim that bodies (since extended) are divisible, but this is slightly slippery since divisibility implies only potential, rather than actual, parts. Perhaps Du Châtelet was sensitive to this in simply asserting that bodies are composites and then asking about the nature of the elements out of which they are composed. See also Stan (2018, p. 482) who suggests (in a footnote) that we turn to Wolff to complete Du Châtelet's theory of composites.

9 Detlefsen (2014, p. 21) writes:

> The PC, in the case of simple beings, establishes both the necessary *existence* and *nature* of simple beings. That is, simple beings are necessarily unextended (their nature), and they necessarily must exist so as to explain the fact of existing composed, extended beings—a fact established by our experience that such composed beings do indeed exist. But this is what we might call a *hypothetical* necessity (as opposed to what we might call an *absolute* necessity). That is, these necessary facts about the simple beings of our world obtain only on the hypothesis that our world actually does exist. God could have not created our world, and had he so chosen to not create any world, then simple beings which actually exist (with the nature of being unextended) need not have existed at all.

10 For a similar statement in Wolff of the interconnectedness of things, see Stan (2018, p. 487).

11 For Du Châtelet, unlike for Leibniz, the law of continuity follows directly from PSR (see Du Châtelet 2009, 1.13). Exactly what Du Châtelet understands by continuity is a complex issue. I have learned a great deal from John Hanson's work on this issue. For a discussion of the origins of Laplacian determinism and its connections to PSR and the law of continuity, see van Strien (2014). I am grateful to Anne Seul for her research on the presence of Du Châtelet in the *Encyclopedia*. See Carboncini (1987), Maglo (2008), and Roe (2018) for Du Châtelet and the *Encylopedia*.

12 For Du Châtelet's engagement with *The Leibniz-Clarke Correspondence* (1998), see Hutton (2012), especially pp. 88 and 89 for the topic of space and time. For the philosophy of physics literature on absolute versus relative space, time, and motion, see Barbour (2001), DiSalle (2006), Maudlin (2012), and references therein. Du Châtelet's place in the historical narrative concerning absolute versus relative space and time arises not only from her *Foundations* directly, but also from the fact that the entries on space and on time in the *Encyclopedia* of Diderot and d'Alembert draw heavily on the *Foundations*.

13 We know that Wolff was an important source for Du Châtelet. We do not know, at the time I am writing, exactly what she read, but Paragraphs 45 and 46 of Wolff's *Rational Thoughts* (see Watkins, 2009, pp. 14–15) are at least similar to materials that she is likely to have had access to and give us an indication of the similarities and differences between her position

and that of Wolff, and of the different use to which she put the resources that she drew from Wolff. Stan (2018) offers a discussion of Du Châtelet's account of bodies in comparison to that of Wolff.

14 My account of Du Châtelet's position agrees with that offered by Stan (2018) in its assignment of a type of idealism to bodies but not to the underlying simples. Stan argues that Du Châtelet's account of simples and bodies is much closer to that of Wolff than of Leibniz.

15 Du Châtelet writes (2009, 6.102):

> So time is really nothing other than the order of successive beings; and one forms the idea of it inasmuch as one considers only the order of their succession. Thus there is no time without true, successive beings arranged in a continuous sequence; and there is time as soon as such beings exist.

Thus, once there are simples, with their continuous sequence of states, there is time, independently of our perceptual experience of those simples. It is also worth noting that Du Châtelet follows Locke (*An Essay Concerning Human Understanding*, Book II, Chapter XIV) in asserting (2009, 6.108–9) that our notion of time arises from the succession of our ideas, not from observed motions of bodies, and in rejecting the equating of time with motion. The role of motion is in the measurement of time, in the establishment of a shared, common time. In the opening sentence of (6.113), Du Châtelet notes the asymmetry in the measurement of time versus space present in Newton's scholium to the definitions (Newton, 1999, pp. 408ff., see Brading, forthcoming), and notes the challenge this poses for establishing a common measure of time. She then writes (6.114), "The only one that might be universal is what is called *an instant*; for all men necessarily know this portion of time, which flows while a single idea stays in our mind".

16 Stan (2018) argues that Wolff's elements stand in causal relations to one another and that they act on one another (pp. 486–7), and argues that Wolff, rather than Leibniz, is the most important source for Du Châtelet's account of simples and of bodies.

17 See Paragraph 603 in Wolff's *Rational Thoughts*, translated in Watkins (2009, p. 42). For Wolff's appeal to our representation of simples in his account of space, see Paragraphs 45 and 46 of his *Rational Thoughts* (Watkins, 2009, pp. 14–15), and in his account of composites, see Paragraph 604 (Watkins, 2009, p. 43). It seems highly likely that Du Châtelet drew on these, or on similar passages from elsewhere in Wolff's writings, in developing her account. For further details and discussion, see Stan (2018). Rutherford (2004) offers a comparison of Leibniz and Wolff on the status of simples and their explanatory role (or lack thereof) with respect to phenomenal experience, and discusses Kant's response to these different positions. Watkins (2006) provides another useful starting point for locating Du Châtelet's simples (their nature, status, relationship to space, and so forth) in comparison with those of Leibniz, Wolff, and Kant.

18 This chapter contains Du Châtelet's accounts of condensation and rarefaction, and of cohesion, which were test cases for any matter theory of the time.

19 For an extended treatment of the debates surrounding actual and potential parts in the 17th and 18th centuries, see Holden (2004).

20 Keill (1720), Lecture III "Of the Divisibility of Magnitude", 's Gravesande (1720) Chapter IV "Of the Divisibility of Body in Infinitum; and of the Subtility of the Particles of Matter", and Musschenbroek (1744) Paras. 23–28 but not in Pemberton (1728). The problem also lies at the heart of Kant's "Physical Monadology" (so-called) of 1756 (see Kant, 1992, pp. 51ff.). His solution is to introduce simples that are located *in* space as points, and each *fills* a determinate finite region of space around its location by means of a "sphere of activity". Unlike Kant, Du Châtelet does not take space as given (her simples are not located in space), but as derivative, and she resists the retreat to "spheres of activity", which Galileo and Descartes had sought to remove as unintelligible (see Henry, 2011), attempting to retain a mechanical approach as an ideal in the theorizing of bodies and their motions (see *Chapter 4*).

21 At the time Du Châtelet was writing, the influence of Malebranche's occasionalism remained strong in France. For discussion of whether Descartes himself was an occasionalist, see Lee (2016, Section 1.3) and references therein.

22 Both the following quotations from the manuscript are found in Janik (1982, p. 101). I thank Lauren Montes for the English translation given here.

23 The text of "On Liberty" is available as Appendix I in Voltaire (1989). Janik (1982, p. 88) identifies "On Liberty" as having been written by Du Châtelet rather than by Voltaire, and as having originally been intended to form Chapter 5 of the *Foundations*. Moreover, Janik notes that passages from Chapter 21 of the manuscript which were omitted in the published version echo arguments from "On Liberty". These passages are to be found beginning on p. 328r of the manuscript (p. 745 of the pdf) available through the BNF. Chapter 21 is the chapter on *vis viva*. See also Iltis (1977, pp. 31–2), who argues that free will was a central issue for Du Châtelet. My thanks to Anne Seul and Lauren Montes, whom I consulted in translating "On Liberty".

24 We know that Du Châtelet studied Locke's *Essay*, perhaps in English, but certainly in Pierre Coste's French translation (1700). Locke considers the two sorts of action of bodies of which he maintains we have any idea: thinking and motion. It seems to me that Locke's famous passage concerning the "powers" that produce these actions (*Essay*, Book II, Chapter XXI, Para. 4, see also Para. 8) provides crucial context for Du Châtelet's account of human liberty and her interest in the powers of bodies to act.

25 Du Châtelet discusses essence, attributes, and modes, in Chapter 3 of the *Foundations*. See *Chapter 2* and references therein.

26 Detlefsen (2014, p. 23) explains the form of reasoning as follows:

> In using the PSR, we observe the phenomenal facts of our actual world, and we can then posit hypotheses about what must further be true of substance in order to make those facts possible. We see this method at work in several sections of the *Institutions*; for example, monads are active, she concludes, because this would explain the brute fact of motion in the phenomenal world, and she goes so far as to claim that force is also necessary to the nature of matter (IP §139).

In a footnote to this passage, Detlefsen draws attention to Gale's (1970, pp. 114–27) labeling of this sort of argument in Leibniz as a "regressive metaphysical argument" and notes a suggestion by Eileen O'Neill that we may read Du Châtelet's use of PSR as providing something like an inference to the best explanation. I find all of these suggestions helpful in thinking about how to further analyze and critique the form of reasoning being used by Du Châtelet.

27 In this paragraph, she switches her terminology from motive force to active force, but I *think* she means the same thing by both. Leibniz used the term "active force". Watkins (in Kant, 2012a, p. 689) writes that Wolff adopted Keill's terminology of "motive force" and used it interchangeably with "active force" in his *Cosmologia generalis*.

28 For Du Châtelet on the problem of change, I have learned much from Aaron Wells.

4 Bodies in Action

The problem of how bodies are capable of acting on one another is the key issue that motivates and drives the arguments of Du Châtelet's *Foundations of Physics*. Chapters 1–4 of the *Foundations* provide the framework within which Du Châtelet develops her account of bodies and their motions and, most importantly in my view, these chapters set out her method for theorizing in physical science. Chapters 5–10 deal with controversial issues concerning space, time, and matter, using resources from Chapters 1–4. Chapters 5 and 6 are on space and time respectively, while Chapters 7–10 provide Du Châtelet's account of the bodies that are the subject matter of physics, such that these bodies are capable of acting on one another. Chapters 11–21 discuss the motions of these bodies, including the laws of motion (Chapters 11 and 12), gravitational motion (Chapters 13–19), contact action (Chapter 20), and *vis viva* (Chapter 21). In what follows, we explore specific examples of bodily action. As we shall see, Du Châtelet makes use of both her two-pronged methodology (for discussion of which see *Chapter 2*) and her account of bodies and forces (see *Chapter 3*) to tackle the most prominent cases of bodily action in physics at the time: collisions, Newtonian gravitation, and *vis viva*.

The paradigmatic case of body-body action, ever since Descartes's *Principles of Philosophy* (published in 1644; see Descartes, 1991), had been two-body collisions. In the 17th and early 18th centuries, those who adhered to a "mechanical philosophy" advocated contact action as the sole means by which bodies act on one another, with collisions being the means by which all change in the world comes about. This view of bodily action was highly influential among philosophers throughout Europe. Yet, at the time that Du Châtelet was writing, almost a century after the publication of Descartes's *Principles*, a causal-explanatory account of collisions remained elusive.[1] I argue that Du Châtelet's attempt to provide such an account is revealing, highlighting the depth of the problem as it stood at the time.

Du Châtelet's account of collisions relies on her account of the relationship between the forces of bodies and their motions, so we begin with this (Section 4.1), before examining her account of collisions (Section 4.2). As we will see in her discussion of collisions, Du Châtelet commits herself to contact action as the sole means by which bodies act upon one another. I examine this commitment to mechanism, suggesting that Du Châtelet lacks the metaphysical resources to justify her commitment, and arguing that her reasons for adopting mechanism are better understood as methodological than metaphysical (Section 4.3).

For Du Châtelet's audience in 1740, both in France and across Europe and the British Isles, two controversial cases concerning bodily action loomed large: Newtonian universal gravitation, understood as involving action-at-a-distance, and the so-called "*vis viva* controversy" over the true measure of the force of a body. I show that Du Châtelet applied her two-pronged methodology to address these controversial cases, and examine her position on each for its reliance upon and consistency with her account of bodies and their forces (see Sections 4.4 and 4.5).[2]

In this chapter, I make clear the continuity of Du Châtelet's project throughout the *Foundations*. This conclusion is contrary to the "Received View" (see *Chapter 1*), according to which the later chapters on "Newtonian physics" are detachable from the opening chapters on "Leibnizian metaphysics". In my view, the *Foundations* is unified by its central concern – the problem of bodily action – and by the methodology that Du Châtelet develops and applies in order to address this problem. The examples discussed in this chapter substantiate this claim. In Section 4.6, I summarize the main conclusions of my study of Du Châtelet's *Foundations*.

4.1 Force and Motion

Du Châtelet begins Chapter 11 by offering a general definition of motion and then distinguishing motion into three kinds: absolute motion, common relative motion, and proper relative motion. She defines absolute motion as motion with respect to other bodies, themselves considered to be immobile, and absolute rest as "the permanence of a body in the same absolute place, this is to say, the continuation of the same relationships of the body being considered with respect to the bodies which surround it, themselves considered to be immobile" (2009, 11.222, translation slightly amended).[3] She does not tell us what these other bodies are that provide the standard of absolute motion and rest, but the fixed stars seem a likely candidate given that she asserts that the Earth is in motion. Such absolute motions are the real motions, but Du Châtelet distinguishes these *definitions* of absolute motion and rest from her account of the *causes* of real motion and rest.[4] The cause of real motion is the force of motion in the moving body.

Du Châtelet's clearest statements of the relation of force to motion and rest are these (2009, 11.223 and 11.225)[5]:

> When the active force or the cause of motion is not in the body which can move, this body is at rest, and this is, strictly speaking, real rest.
>
> [T]he only real motion is that which operates by a force residing in the body that moves, and the only real rest is the absence of this force.

By way of illustration, Du Châtelet appeals to an example of a stone thrown while on a ship. If the stone is thrown in one direction with a given speed, while the thrower stands on the ship which is moving in the opposite direction at the same speed, the stone will be seen by those on the shore to remain at absolute rest with respect to the horizontal direction, she says (2009, 11.218). The reason for this is that, in moving with the ship, the stone acquired a force in the direction of motion of the ship that is exactly equal to the force which it acquires in being thrown in the opposite direction. The upshot is that the stone remains at absolute rest in the horizontal direction. And the reason for this, in turn, is that the "two equal and opposite forces mutually destroy one another" (11.218, my translation).

Du Châtelet uses the principle of sufficient reason (PSR) to conclude that a body in motion will continue to move (2009, 11.227 and 228). A moving body has associated with it a quantity of active force, by which it moves at a certain speed (and presumably, though see later, in a certain direction): any change to that speed requires a change in the quantity of active force, and such changes cannot happen without a reason. A body at rest is without active force, but this does not mean that it lacks any force whatsoever. On the contrary, as we saw in *Chapter 3*, an essential property of matter is passive force, by which a body resists motion. Appealing to PSR once again, Du Châtelet claims that in order for a resting body to begin to move, there must be some cause. For Du Châtelet, these two appeals to PSR provide the basis of Newton's first law of motion (Newton, 1999, pp. 416–17), for which she gives a subtly but importantly different statement (2009, 11.229, amended translation):

> A body perseveres in the state it is in, be it rest or motion, unless some cause removes its motion or its rest.

Unlike Newton, Du Châtelet does not specify that the preserved state of motion is at a constant speed in a straight line. I suspect that she may have taken this to follow from the unchanging quantity of active force that maintains the body in its state of motion. However, once that quantity is measured by mv^2 (as she had come to believe, in endorsing *vis viva*; see later) rather than by mv (as she originally believed, when following the standard Newtonian position), it is no longer a directional quantity. That to one side, the more important difference for our purposes is the explicit appeal to causes in the statement of the law. Newton formulates his law in terms of impressed forces, and he is explicit that these can be treated mathematically, whereas Du Châtelet appeals directly to causes, and deliberately collapses the methodological divisions set up so carefully by Newton in the *Principia*.[6]

We will not pause here to assess the strength of these arguments for a body preserving its state of rest or motion. Our interest is in outlining Du Châtelet's position. Du Châtelet defines the motion and rest of a body relative to other bodies, and then provides a causal account of this motion and rest in terms of the active and passive forces of bodies previously introduced (in Chapter 7). Given these active and passive forces and their connections to rest and motion, Du Châtelet then uses PSR to argue that bodies preserve their states of rest and motion. This, for Du Châtelet, is the basis of her version of Newton's first law of motion. With this in place, we can now turn our attention to her account of collisions.

4.2 Collisions

How do bodies change their states of motion? Without argumentation, Du Châtelet simply states (2009, 11.229), "The active force and the passive force of Bodies is modified in their collision, according to certain laws that can be reduced to three principles". And she then states her version of Newton's laws. We have already seen her version of Newton's first law. Her version of the second and third laws is as follows (2009, 11.229)[7]:

> Second Law: The change that happens in the motion of a body is always proportional to the motive force that acts on it; and no change can happen to the speed and the direction of the moving body except by an exterior force; for without that, this change would happen without sufficient reason.
>
> Third Law: The reaction is always equal to the action; for a body could not act on another body if this other body did not resist it. Thus the action and the reaction are always equal and opposite.

Du Châtelet argues for both these laws by applying PSR within the context of her account of forces.[8] The contrast with Newton's approach is clear. Newton begins by treating force mathematically, leaving open both the physical process by which the motions of bodies are changed in accordance with his laws, and whether impressed forces are contact forces or not. Du Châtelet seeks to justify her laws of motion within the context of her prior theory of bodies and forces, and by means of PSR. Moreover, she commits herself to contact action, and to collisions as the sole cause of changes in the motions of bodies (see Section 4.3).

Differences in wording from Newton's formulations notwithstanding, if we allow that Du Châtelet has Newton's laws, then she can solve whatever quantitative problems these laws can be used to solve. Therefore, she can solve all the collision problems soluble with the resources of Newton's laws of motion, and she can claim comparable empirical support for her laws insofar as they correspond to those of Newton.

However, for Du Châtelet, the mathematical solution of collision problems via laws of motion does not provide a complete solution to the physical problem of collisions. The mathematical solution provides a rule for calculating the outgoing speeds and directions of colliding bodies, given their incoming speeds and directions (for example), but it fails to provide a causal account of how the bodies act on one another: it is incomplete as a physics of collisions.

At the end of Chapter 11, Du Châtelet introduces additional resources (2009, 11.268):

> When a moving body encounters an obstacle, it strives to displace this obstacle; if this effort is destroyed by an invincible resistance, the force of this body is a *force morte* [dead force], that is to say, it does not produce any effect, but it only tends to produce one.
>
> If the resistance is not invincible, the force then is *force vive* [live force], for it produces a real effect, and this effect is called *the effect of the force of this body*.

In *Chapter 3*, we saw that all bodies have associated with them an active force, by which they move (or strive to move), and a passive force, by which they resist motion (or changes in motion).[9] In the previous passage, Du Châtelet explains that the active force of a body manifests itself in two ways: as dead force (*force morte*), when the body strives to move but fails (due to an obstacle), and as living force (*force vive*),[10] when the body is in motion. Du Châtelet further develops her account of the forces of bodies in Chapter 20, and this enables a causal explanation of the process of collisions, as follows. During the time when colliding bodies are in contact they press upon one another by means of dead force. In so doing, they impress potential speed, or a tendency to motion, into one another. Eventually, all the active force of one body is used up, and that body can no longer actively resist the pressure of the other through its own dead force. The other body continues to impress active force on the first body but now, meeting no active resistance, this impressed force becomes a living force. The upshot is that the first body begins to move back in the direction of the impressed force, with an acceleration that depends upon its quantity of passive force (i.e. its inertia). If it happens that the active forces are equal, then they destroy each other and the bodies remain at rest. If it happens that the two bodies are moving initially in the same direction, the dead force impressed by the faster body on the slower becomes a living force, and the slower body increases its speed. If it happens that the line of impact is oblique, the outcome of the collision depends upon the components of force acting along the line of impact. This is an entirely qualitative causal-explanatory story that has to be combined with her version of Newton's laws of motion in order to arrive at a quantitative account of collisions.[11]

Du Châtelet's belief that providing the laws of motion, or more specifically some rules of collision, is insufficient as an answer to the question "how do bodies act on one another?" is in contrast with the view of 's Gravesande, for example, who maintained that once we have provided the rules by which bodies move, we have gone as far as we can in providing the causes of those motions.[12] For Du Châtelet, providing only the rules does not suffice to provide a physics, for it fails to give an account of the causes. Where 's Gravesande saw an *epistemic* limitation, believing laws to be discoverable in the phenomena but their underlying causes to lie beyond our reach, Du Châtelet saw a limitation of the *method*. She developed a method intended to overcome the problem and provide epistemic access to underlying causes.

The upshot is an integration of a broadly Leibnizian approach to forces with broadly Newtonian laws of motion, a task more complex than might at first appear. For Newton, there is no dynamical distinction between rest and uniform motion, and Newton eliminates the distinction between agent and patient (the distinction becomes a matter of how we describe an interaction, rather than reflecting anything in the dynamics of the interaction itself). Du Châtelet seeks to recover the motions of bodies, as captured by Newton's laws of motion, whilst retaining a distinction between rest and motion in terms of the forces of a body, and relatedly between agent and patient, in her causal account of how bodies act on one another. This is either a brilliant integration of the two approaches (insofar as it succeeds) or a spectacular failure to appreciate one of the deepest philosophical insights of the *Principia*.

As a first step in evaluating Du Châtelet's position, we can examine the extent to which the account of collisions in terms of dead and living force succeeds in providing a causal explanation. At the time Du Châtelet was writing, there were three metaphysical accounts of body-body causation available: occasionalism, preestablished harmony, and physical influx.[13] Du Châtelet rejected the first two as failing to provide genuine causal interaction among bodies, and instead developed an account that avoids the main objections against physical influx. Leibniz had rejected physical influx on the following grounds. If change is the acquisition (or loss) of an accident, then in order for one substance to give rise to changes in another substance, it must be responsible for a new accident in that second substance. This requires either that the accident literally migrate from one substance to another or that the first substance create a new accident in the second substance out of nothing. Leibniz rejected both: the first since accidents have no independent being, and the second on the grounds that it involves creation *ex nihilo*, which he rejected as a power inappropriate for created substances.

In many places, Du Châtelet operates with a substance notion of force, wherein force is "communicated" from one body to another, and "consumed" by resisting bodies (see especially 11.231–4 and 11.259). At times, she also adopts a "contest" interpretation of collisions, which is one way to read Descartes on collisions (see Garber, 1992, pp. 234–5, and Gueroult, 1980), writing (2009, 11.259):

> In any action, the body that acts, and that against which it acts, fight each other, and without this kind of fight there can be no action; for I ask how a force can act against that which does not offer any resistance.

I think that in order to make sense of her account of body-body causation, we need to appeal to her account of essence, attributes, and modes, and that this further explains and justifies the role of Chapter 3 in the *Foundations*. For Du Châtelet, the attributes of a thing follow necessarily from its essence, but its modes are not determined by its essence. Rather, from its essence follows a range of ways in which a thing, in order to be actualized, must be determined, but the particular way in which it is actualized remains undetermined by the essence. For example, it follows from the essence of body that it has shape, since extension belongs to its essence, but the *particular* shape of a body is *not* determined by the essence. Rather, it is determined at least in part by the prior state of the body, and perhaps also in part by its causal interactions with other bodies. Du Châtelet's simples are causally interconnected, and her bodies, though phenomenal, are also causally interactive in virtue of the causal interconnections of the simples out of which they arise. Extension, active force, and passive force are the essential properties of a body, but quantity of extension, quantity of active force, and quantity of passive force are modes: they can vary without the thing ceasing to be a body. It seems to me that in terms of metaphysical resources, Du Châtelet is in a promising position for offering an account of body-body causation. Her account, in terms of the limitations of modes, overcomes Leibniz's objection to physical influx.

While the account may be promising metaphysically, it faces a serious problem from the perspective of Du Châtelet's own methodological requirements. The "communication" of force from one body to another, the contest, the fight, the transfer, the using up and consuming of force: all these words are labels whose imagined referent is empirically inaccessible. Indeed, while occasionalism, pre-established harmony, physical influx, and Du Châtelet's variable mode theory tell very different causal stories metaphysically, Du Châtelet offers no resources for distinguishing them empirically. Her two-pronged methodology demands that a hypothesis not only satisfy the principles of our knowledge (such as PSR), but also that it be empirically testable. Therefore, judged by her own standards, the causal explanation of body-body collisions offered in her account should be assessed as nothing more than an attractive fiction.

We can summarize her achievements as follows. In Chapters 7–9 of the *Foundations*, Du Châtelet provides an account of how it is possible for there to be bodies acting on one another, in terms of extension, active force, and passive force. These latter forces are derivative forces: they derive from the primitive forces of the simples out of which bodies arise. In Chapter 11, Du Châtelet turns her attention to the causal explanation of the phenomena of bodies in motion in terms of the attributes of bodies: extension and derivative force. Our question was whether she is able to offer a causal explanation of collisions. We can break our assessment into two steps. First, her account shows how it is possible for bodies to act on one another such that the problem of collisions can be addressed within that account. Insofar as Du Châtelet's statement of the laws of motion is sufficient to solve the problem of collisions, and insofar as the concept of force to which the laws of motion appeal is the same as the one introduced to solve the problem of bodily action, Du Châtelet addresses both these issues with a single concept of force. This is no small achievement. However, Du Châtelet's own two-pronged method requires a further step, and here she runs into difficulties with empirical accessibility: her own causal story seems to be empirically indistinguishable from all the other available accounts. Thus, even were we to grant that she has succeeded in showing how it is possible for bodies to act causally on one another in collisions, she has failed to show that this is actually how they act on one another. To do so requires empirical access to the process by which they act, and this is not available.

4.3 Mechanism and Method

The account of force and bodily action offered so far makes a commitment to contact action, seemingly without argument. Moreover, Du Châtelet repeatedly asserts her commitment to mechanism at different points throughout the text, and this commitment plays a crucial role in her discussion of Newtonian gravitation. So, before turning our attention to gravitation, we should take some time to track back, and to look at her arguments for adopting a mechanical philosophy.[14] The first takes place in one short paragraph in Chapter 8, which reads as follows (Du Châtelet, 2018, 8.140, my numbering):

(1) The first thing we understand about bodies is that they are Beings composed of several parts.
(2) Thus, the properties of a composed being must be suitable for them;
(3) however, there cannot be changes in the composite except in terms of its figure, its size, the situation of its parts, and the place of the whole:
(4) and consequently all the changes in bodies must reduce themselves to these things.
(5a) But as none of these changes come to pass without motion, (5b) all changes must be caused by the motion of matter, or from that which is extended.
(6) Thus, all bodies, all portions of matter, are machines;
(7) for we call machine a composite for which the changes are in virtue of its composition and through the means of motion.

The tricky premise, when it comes to Newtonian gravitation, is premise (5). Here, Du Châtelet moves from the claim (5a) that "none of these changes come to pass without motion" to the claim (5b) that "all changes must be caused by the motion of matter". But whereas (5a) seems to follow from (1) to (4), (5b) is more problematic. In particular, when it comes to the changes in the "place of the whole" (see (3)), it seems especially clear that (1)–(5a) do not yield (5b) without further assumptions. Yet changes in the "place of the whole" are precisely what is at stake in Newtonian gravitation: when two planets move toward one another due to mutual attraction, there is a change in the "place of the whole" for each planet. Claim (5b) is problematic because it invokes the *causes* of the changes described in (1)–(4). The assumption required to make the inferences from (1)–(4) to (5b) valid is that we read (4) as including causes: the causes of the changes (rather than just the changes themselves) must be reduced to figure, size, situation of parts, and place of the whole. But this is simply to deny the Newtonians their claim of universal gravitation as a property of matter, rather than to offer an argument against it.

A second argument occurs in Paragraphs (8.145–6). In Paragraph (8.145), Du Châtelet argues that extension, active force, and passive force are sufficient to explain all the changes that occur in bodies. Bodies have a size, a shape, and a situation in virtue of being extended. Changes in their size, shape, and situation are possible in virtue of their active force. Which of these possible changes become actual depends on the passive or resisting force of bodies. She then writes (2018, 8.146):

> We see from this that the Philosophers who would like to admit into Philosophy only mechanical principles, and who claim that all natural effects must be explained mechanically, are right; for the possibility of an effect must be proven by the shape, the size, and the situation of the composite, and its actuality by the motion. And whoever reasons in this way proceeds in their reasonings as the nature of things requires.

The move from her account of the active and passive forces of bodies to endorsing mechanism depends upon (a) the means by which the active force, by which one body strives to move another, can be "felt" by the other body, and (b) the means by which that body can, in turn, resist the action of the first body via its passive force. Du Châtelet's assumption seems to be that these can occur only in contact between the two bodies. The intuition seems to be that bodies "feel" the presence of one another via their mutual impenetrability, which follows (Du Châtelet assumes) from their extension. But the claim that impenetrability is the only means by which bodies "feel" the presence of one another is never argued for in Du Châtelet's text, and it is an assumption that proponents of action-at-a-distance reject.

We might instead attempt to construct an argument for mechanism from Du Châtelet's account of the causal relations among simples, and the relationship of these to the causal relations among bodies. According to Du Châtelet, all simples are causally interconnected with one another, with these causal relations spreading out in diminishing ripples from each simple. In other words, the simples are *causally ordered*, though *not* (as Du Châtelet makes clear) spatially ordered. The causal relations of bodies supervene on this underlying nonspatial causal order among the simples: bodies, *and the spatial relations among them*, are our representation of the simples (see *Chapter 3*). However, further argument would be required to show that these spatial relations reflect the underlying causal order of the simples in such a way that contact action is the only causal relation among the bodies. Nothing in Du Châtelet's account of causal ordering among the simples rules out action-at-a-distance in our spatial representation of that causal ordering.[15]

We have two observations in hand. First, Du Châtelet consistently asserts a commitment to mechanism. Second, from the perspective of her metaphysics, she provides no convincing argument to support this commitment, and it seems that she lacks the resources to do so. This brings us to my third point: Du Châtelet's commitment to mechanism plays a primarily *methodological* role (see also *Chapter 2*, where I argue that Du Châtelet's *Foundations* should be read as being framed by method).

In Chapter 8, Du Châtelet insists on mechanical explanations as the goal, writing (2018, 8.162), "For we must try, as far as possible, to explain the Phenomena mechanically, that is to say, by matter and motion". However, her commitment to mechanism is nuanced, reflecting its utility in theorizing. Du Châtelet chastises those who rush prematurely to a mechanical explanation, arguing that we must proceed with caution (9.182). Indeed, the account of matter and body that Du Châtelet offers in Chapter 9 has the consequence that mechanical explanations will often be epistemically out of reach. For Du Châtelet, matter is divisible – and divided – so far beyond the limits of our senses that we cannot hope to discover the shapes, arrangements, and motions of these smallest parts of matter. Among the sensible phenomena are physical qualities, such as magnetism, electricity, fire, cohesion, elasticity, and gravity, that are most likely produced by a matter so fine and fast-moving as to be beyond the reach of our senses. This has implications for our investigations and theorizing (2014, 9.176):

> So it is a waste of time to attempt to divine these imperceptible mysteries, and we should limit ourselves to observing carefully the qualities that fall under our senses and the phenomena that result, which we can employ to make sense of other phenomena that depend on them.

Discovering these observable qualities is the first step in physics. Du Châtelet insists that we should not rush to attempt mechanical explanations where those are out of the reach of empirical investigation, for "these physical qualities, which make up the effect of mechanical causes, must necessarily precede them in the explanation of phenomena" (2014, 9.184).

Despite this, Du Châtelet is adamant that we should not give up on our pursuit of mechanical explanations. In the wake of Newton's *Principia*, Newtonian philosophers had appealed to attraction to explain a wide variety of phenomena, but Du Châtelet worried that this style of explanation allowed philosophers to stop too soon in their research into the phenomena (this being bad for the progress of science) and to rest satisfied with explanations that fail to meet her standards of intelligibility (see later). Her remedy was to insist that the explanatory quest terminate not with attraction, but with mechanism.[16] She suggests (2018, 8.163) that we should make use of all the "physical qualities" (such as "the elasticity of air, the fluidity of water, the heat of fire"), even though we do not yet have a mechanical explanation for them, in order to provide interim explanations "for other properties that occur together in nature and that originate from a mixture of some of them". Proceeding one step at a time, we use simpler physical qualities to explain more complex ones. Once we have moved down through layers of physical qualities, we will at last be in a position to seek mechanical explanations of those simpler qualities. Meanwhile, we should not mistake nonmechanical explanations for ultimate explanations, which are always mechanical. Du Châtelet writes (2014, 9.182):

> however difficult it may be to apply mechanical principles to physical effects, one must never abandon this manner of philosophizing, which is the only good one, because it is the only one with which one can make sense of the phenomena in an intelligible fashion.

Insofar as that which is intelligible is that which satisfies the principles of our knowledge, Du Châtelet's position seems to be that only mechanical explanations satisfy PSR.[17] Mechanical explanations are an explanatory ideal, a goal, that should be approached gradually, and under the constraints of her methodology (see *Chapter 2*).

We must seek mechanical explanations, and in so doing we further physical science. For Du Châtelet, it is a methodological requirement that we not resort prematurely to God, and specifically to the arbitrary will of God; she equates stopping at nonmechanical explanations with invoking the arbitrary will of God.[18] Where we cannot find a mechanical explanation, we must admit our ignorance, for appeal to the will of God does not advance physical science (see 2018, 8.162). In practice, we will not always be able to provide mechanical explanations. This is both epistemically inevitable and methodologically admissible.

4.4 Gravitation

Du Châtelet used her two-pronged methodology to intervene in the controversy between Newtonian universal gravitation, considered as an action-at-a-distance theory, and the alternative approach favored in France of a vortex theory. In Newtonian universal gravitation, every particle acts on every other particle in the universe with a force in accordance with Newton's law of universal gravitation. Those who opposed Newtonian universal gravitation argued that action-at-a-distance is unintelligible: how could a body act where it is not? They accused Newton of reintroducing mysterious powers into our account of the natural world, after such figures as Galileo and Descartes had worked so hard to put physics on a new footing, free of such mysteries. Descartes had proposed a vortex theory of planetary motion, according to which all the planets are swept around in a plenum[19] of fine matter, like a stone in a whirlpool, so that collisions between the planets and the particles of the vortex are all that we need in order to explain the gravitational behavior of the planets. Descartes had also proposed a plenum account of the gravitational behavior of terrestrial bodies, such as the fall of stones. In the plenum account of gravitation, it is the body as a whole (the planet or the stone, in our examples) that is being affected by the surrounding plenum. This is in contrast to Newton's account, where each particle belonging to the body acts on every other particle (both in the same body and in all other bodies). In his *Principia*, Newton (1999) argued that vortex theory failed to explain the observable phenomena, and offered his alternative account: universal gravitation.

The problem addressed by Du Châtelet is this: how are we to decide between these two theories? One might think that the issue is already decided, given her commitment to mechanism in her account of bodily action. However, this is not the case. Instead, Du Châtelet applies the method set out in the earlier chapters of the *Foundations*, and seeks a conclusion accordingly. She presents detailed arguments drawing on both empirical evidence (Chapter 15) and the principles of our knowledge (Chapter 16). Du Châtelet thus deploys both prongs of her methodology in order to address the problem of how we decide between the two alternative theories.

In Chapter 15, Du Châtelet introduces Newtonian gravitation and Huygens's version of vortex theory. She then turns to the empirical evidence and considers two arguments.[20] The first argument concerns the planetary trajectories. In Book 2 of the *Principia*, Newton had argued that if the matter making up the vortex is of the same kind as the matter making up the planets, and is therefore subject to Newton's laws of motion, then "the hypothesis of vortices can in no way be reconciled with astronomical phenomena" (Newton, 1999, Book 2, Section 9, Scholium to Proposition 53). Huygens responded by rejecting the idealizations and assumptions about fluids that Newton used in making the argument go through. He offered instead a vortex theory recovering the trajectories of the planets (Huygens, 1690). Supposing this successful, the upshot is that empirical evidence does not distinguish between Newtonian universal gravitation and vortex theory for planetary trajectories.[21]

Du Châtelet then moves on to a second argument, concerning the shape of the Earth. This is something that she was very interested in, corresponding frequently with Maupertuis.[22] She notes that the two approaches, Newtonian universal gravitation and Huygens's vortex theory, give rise to *different* predictions in this case. She writes (2018, 15.379):

> Mr. Huygens believing to be the same everywhere [because it pertains to the body considered as a whole], and Mr. Newton assuming it to be different in different places on the earth, and dependent upon the mutual attraction of the parts of matter: the only difference to be found in the shape that these two Philosophers attributed to the earth was that from Mr. Newton's theory there resulted a greater flattening than from that of Mr. Huygens.

So she is very clear about the difference between the two approaches being due to the disagreement over universal gravitation (i.e. whether it is particle to particle or not), and on where the observational consequences differ. She is also up to date with the efforts to measure the shape of the Earth, and reports that she is awaiting further results that will help determine the question between Huygens and Newton. In short, by the 1730s, the empirical evidence on the shape of the Earth favored Newtonian universal gravitation.

However, Du Châtelet does not accept the empirical evidence against vortex theory as conclusive. This is, at least in part, because we have so far considered only one prong of the methodology by which hypotheses and theories are to be tested. The other prong of her methodology draws on the principles of our knowledge. The argument against Newtonian universal gravitation on the basis of PSR takes place primarily in Chapter 16.[23] The chapter begins with Du Châtelet arguing against the overuse of attraction for explaining phenomena.[24] Her commitment to mechanical philosophy is intended as a correction to this, by requiring that even where attractions are posited, this should be viewed as temporary and a tool toward seeking an underlying mechanical cause (see Section 4.3). Next, Du Châtelet turns her attention to gravitational attraction. Unlike Maupertuis, who considered the sufficient reason for the *particular* form of the law of attraction (i.e. for the *inverse square* law, see Maupertuis, 1732, and also Terrall, 2002, pp. 79–81), Du Châtelet is concerned with the sufficient reason for attraction itself (2018, 16.393–4). That is, she questions the sufficient reason for the existence of *any* law of attraction and argues that *no* such law can satisfy PSR.

Du Châtelet's strategy for showing the inadmissibility of Newtonian attraction as a cause of gravitational phenomena is to imagine a universe that is "void by supposition" and then to show that, in this case, there can be no sufficient reason for bodies in that universe to obey any law of attraction.[25] In the opening scenario, there are two spatially separated bodies which, by supposition of Newton's law of gravitation, begin to move toward one another, each thereby undergoing a change in its state of motion. By PSR, there must be a reason for this change. By assumption, the reason for the change in any one of the bodies must be either internal to that body or external to it. As was argued in Chapter 10, a body requires an external cause in order to change its state of motion, so an internal cause is ruled out. The only available external cause is the other body. However, by stipulation, there is a void space between the two bodies, and this rules out anything traveling between the two bodies such that the one could cause a change in the other. Thus, Du Châtelet concludes, Newtonian gravitation is in conflict with PSR and should be rejected. With the argument set out thus, the problem seems to be that Du Châtelet rules out action-at-a-distance by fiat when she insists that something must travel from one body to the other in order for the first to cause a change in the second. I do not know how to recover her argument from this criticism. She offers a second argument, in which she asserts that not even God could know which way a body will move on the basis only of the properties of the bodies themselves. In this case, she seems to be ruling out the kinds of relational properties on which an action-at-a-distance interpretation of Newtonian gravitation rests. No justification for this is provided, and it is not clear to me whether her account of the attributes of bodies or the forces of bodies (see *Chapter 3*) allows one to be provided. In short, the arguments given in Chapter 16 seem to me to be not very good.[26] For our purposes here, we can move on to the conclusion that she draws: the principles of our knowledge, and specifically PSR, tell against the admissibility of the Newtonian theory of gravitation.

We can summarize the overall argument concerning gravitation, and its upshot, as follows. In Chapter 15, Du Châtelet looks in detail first at planetary trajectories, concluding that there is insufficient empirical evidence to decide between the two theories, and second at the shape of the Earth, concluding that the empirical evidence favors Newton's theory. In Chapter 16, she argues against Newtonian universal gravitation on the basis of PSR. The outcome of deploying her two-pronged methodology is that the empirical evidence favors Newtonian universal gravitation whereas the principles of our knowledge favor vortex theory. Thus, at the time she was writing, the upshot was inconclusive, and Du Châtelet finished her chapter on Newtonian attraction accordingly, suggesting that (2018, 16.399):

> perhaps a time will come when we will explain in detail the directions, motions, and combinations of fluids that bring about the Phenomena that the Newtonians explain by attraction, and this is a quest to which all Physicists must apply themselves.

This might seem like a rather disappointing result, but I think Du Châtelet's conclusion is entirely appropriate. Du Châtelet has successfully demonstrated the power of her methodology for driving theorizing forwards: she has pinpointed precisely where the differences between the theories or hypotheses currently available lay, both in terms of empirical evidence and in terms of the principles of our knowledge, and she has made explicit what each theory needs to address in order for the debate to move forward. Understood in this way, the situation at the time was indeed inconclusive. Even though Du

Châtelet's account of bodily action and commitment to mechanism favored vortex theory over Newtonian gravitation, Du Châtelet stuck to her own methodological prescription and withheld final judgment.

4.5 *Vis Viva*

The final example of bodily action that we will consider arises in the so-called "*vis viva* controversy", reignited in no small part by Du Châtelet herself, prior to the publication of the *Foundations*.[27] The dispute concerned whether the proper measure of the motive force of a body depends on its speed (or later velocity) or on the square of its speed.[28] Once again, Du Châtelet mobilizes both prongs of her methodology – the principles of our knowledge and empirical evidence – in order to address the problem. The main chapter concerned with this issue is Chapter 21, and Du Châtelet begins this chapter with an argument founded on PSR and the law of continuity. She writes (2018, 21.557):

> You saw in Chapter 1 that the principle of continuity, founded on that of sufficient reason, admits of no exception in nature, and that a body cannot pass from one state to another without passing through all the degrees that lie in between. By this law a body that is at rest cannot suddenly pass into motion. It must go successively, and as it were by degrees, in acquiring one by one all the degrees of movement that are between rest and the movement that the body must acquire.

She goes on (2018, 21.558):

> A body in motion has a certain force that increases and decreases with the speed of the body. So once it is seen that a body does not acquire its total speed all at once, but gradually, it is clear that the force accompanying this speed also passes successively from the cause to the body it puts into motion.

She then proceeds to theorize dead and living forces within this constraint (2018 and 2009, 21.559–75) and thereby to argue for living force (*force vive*, *vis viva*), empirically accessible via mv^2 (where m is the mass and v the speed of the body in question), as the measure of the motive force of bodies. The details of her argument need not concern us. For present purposes, the point is that the first prong of her methodology is deployed to argue for living force. Having done this, she turns her attention to empirical evidence, and from (21.576) onwards addresses arguments for and against the different views on the true measure of motive force, concluding that the evidence favors *vis viva*. Du Châtelet concludes (2018, 21.588):

> We have seen in this chapter that all experiments concur in proving that there are living forces – and metaphysics speaks almost as strongly in their favor as physics.[29]

Thus, unlike in the gravitational case, both prongs of her methodology lead to the same result, and Du Châtelet maintains that she has resolved the issue in favor of *vis viva*. Moreover, it seems that this conclusion is consistent with her account of bodies and forces.

It seems to me, however, that this conclusion is premature. My concern is that her response to one of the challenges to *vis viva* directly undermines her causal explanation of the collision process.[30] The challenge is as follows (see 2009, 21.583). Suppose that two bodies hit each other with speeds that are inversely proportional to their masses. For example, let Body A have mass 1 and speed 3, while Body B has mass 3 and speed 1. Given the account of collisions described earlier, and adopting *force vive* (mv^2) as the measure of the motive force of a body, we have that Body A has a motive force of 9, while Body B has a motive force of 3. She writes, "As first it would seem as if the body which has the most speed, having the most force, according to the doctrine of *forces vives*, must push the other before it". Certainly it seems so, so how can she avoid this conclusion?

The argument she gives is this. Suppose that we have a compressed spring, with one end resting against Body A and the other against Body B. When the spring is released, the speeds of the bodies will be in inverse proportion to the masses, or so she claims: $v_A/v_B = m_B/m_A$. This is accepted by the opponents of *vis viva*, so we shall accept it too. As a result, $m_A . v_A = m_B . v_B = 3$, and were we to use mv as the measure of force, we would conclude that the force of the bodies is equal, and this explains why they come to rest. However, Du Châtelet goes on to argue that the motive forces of A and B are not, in fact, equal. Reversing the direction of motion of the bodies in this scenario, Bodies A and B now approach one another with the initial speeds $v_A = 3$ and $v_B = 1$. Now, Du Châtelet uses the claim (to put the point anachronistically) that in the center-of-mass frame in which A and B come to mutual rest, the ratio of distances traveled is as the speeds: s_A:s_B is 3:1. Then, she uses the claim that the force required to compress the spring is proportional to the distance through which it is compressed, so that the force of Body A is proportional to s_A while the force of Body B is proportional to s_B. This shows that the forces of A and B are not equal, but in the ratio of 3:1, even though they come to rest. Moreover, this concurs with the result that the *forces vives* of A and B are in the ratio of 3:1. It seems, therefore, that Body B is able to consume (to use Du Châtelet's own word) more force than Body A, and this is why they come to mutual rest despite having hit one another with different quantities of force. This is Du Châtelet's response to the challenge.

The problem arises when we return to Du Châtelet's causal-explanatory account of collisions, in which qualitative predictions are rendered quantitative by connecting the force of a body in the causal account to the measurable quantities appealed to in Newton's laws (see Section 4.2). Du Châtelet's escape from the aforementioned challenge robs her account of any predictive power whatsoever: in her new account, the forces of the bodies in collision tell us nothing about the outcomes of that collision unless we also know the capacity of those bodies to consume force, and she has no story to tell about this. In making predictions, all the work is being done by Newton's laws of motion in conjunction with empirically determined coefficients of restitution, and we remain without a causal explanation for either these coefficients or the process of collision. In other words, the move that Du Châtelet makes in responding to the aforementioned challenge to *vis viva* undermines both her causal story of the collision process and the power of the unification that she attempted between her account of the forces of bodies and the laws of motion.

The significance of this troubling conclusion needs to be correctly understood. Du Châtelet made important innovations and contributions in pursuit of a viable account of how it is that bodies are capable of acting on one another: she provided a fourth option for a metaphysics of bodily action beyond the three standard ones; she provided a qualitative account of the collision process that, in conjunction with her metaphysics of bodily action, she believed made possible causal action by bodies; she explicitly connected her causal account to Newton's laws of motion; and she insisted that the account be empirically engaged. If our final assessment is that her account nevertheless fails, then this serves only to highlight the extreme difficulties facing any attempt to give an account of bodily action at that time.

4.6 Conclusions

Du Châtelet opens the *Foundations* by presenting her scientific methodology (see *Chapter 2*). She turned to this method in order to address problems of central concern to her, most especially the problem of how it is that bodies act on one another (see *Chapter 3*). This problem is manifest in attempts to give a causal explanation of the collision process for bodies, and in the controversial issues surrounding gravitation and *vis viva* (see *Chapter 4*). The adoption of the new methodology drives the revisions to the manuscript that occurred shortly before publication, including the introduction of the new metaphysics. This is why I believe that her scientific methodology plays a unifying role with respect to the *Foundations* considered as a whole, and is crucial to our understanding of the text (see *Chapter 1*).

In my view, Du Châtelet saw clearly the most pressing foundational problems in physical science that were left unsolved in the early decades of the 18th century, in the wake of Newton's *Principia*. These problems concern the metaphysics and physics of bodies and forces, and the epistemological and methodological challenges facing the investigation of such things. Hers is the most perspicuous diagnosis, and the most thorough-going attempt to resolve the problems, that I have read.

Notes

1 The widespread problems in treating collisions are one thread in a collaborative research project that I am working on with Marius Stan on matter theory and mechanics in the early 18th century.

2 Hecht (2012, pp. 71–2) presents another example of a disagreement in physics that Du Châtelet sought to resolve, this time involving Galileo and Huygens on the one hand, and Varignon on the other (see Chapter 17 of the *Foundations*).

3 Quotations are from the 1740 first edition of Du Châtelet's *Foundations of Physics*, translated from the original French into English. A complete English translation is available via three sources. See Zinsser and Bour (2009) for the Preface and Chapters 1, 2, 4, 6, 7 (partial), 11 (partial), and 21 (partial). See Patton (2014) for Chapter 9. For the remaining chapters and passages, as translated by Brading et al., see Du Châtelet (2018).

4 It may be that in the manuscript versions of this chapter (which do not survive), absolute rest and motion were defined in accordance with Newton, as being with respect to absolute space. However, in Chapters 5 and 6 of the published version of the *Foundations*, Du Châtelet rejects absolute space and time, adopting instead the relative space and time advocated by Leibniz in *The Leibniz-Clarke Correspondence* (with which she was familiar, see Hutton, 2012). In the definitions of absolute rest and motion offered here, Du Châtelet draws on Descartes's "proper" definition of motion in his *Principles of Philosophy* (1991), Part II, Paragraph 25, calling on bodies "themselves considered to be at rest" as providing the standard for absolute rest and motion. Du Châtelet's terminology of common relative motion and proper relative motion seems to be taken from Musschenbroek (see his 1734 and 1739, or his 1744 for a version in English), though deployed differently.

5 Given her account of force from earlier, it is not surprising that she immediately clarifies this by noting that all particles of matter are always in motion, so that even though a composite body may be at absolute rest, internally its parts will always be in motion (2009, 11.225). So far, so good, for consistency with the account of force developed earlier.

6 Du Châtelet was not alone in reformulating the first law in ways that eliminated Newton's distinctions. In fact, the law was widely reformulated by Newton's followers. A prominent example is Euler's version of Newton's first law in his *Mechanica* (1736), in which he replaces Newton's "impressed force" with "external cause". See Suisky (2012) for a comparison of Euler and Du Châtelet.

7 I have changed "motor" to "motive" in the translation of the second law.

8 For example, Du Châtelet explicates the third law as follows (2009, 11.259):

> This resistance that all bodies present when one wants to change their current state is the foundation of the Third Law of motion, by which the reaction is always equal to the action.
>
> The establishment of this law was necessary so that bodies might act on one another, and that motion, being once produced in the universe, might be communicated from one body to another with sufficient reason.

For a discussion of Du Châtelet's treatment of Newton's laws of motion, see Reichenberger (2018).

9 Recall from *Chapter 3* that these are, in fact, derivative forces, deriving from the active and passive forces of the simples from which bodies arise, but we need not concern ourselves with that for our purposes here.

10 The measure of *force vive* is the topic of the infamous *vis viva* controversy, to which Du Châtelet made prominent contributions; see Section 4.4.

11 Du Châtelet is offering an account of what happens during the process of collision, rather than simply offering a mathematical rule that relates final to initial states. Suisky (2012, p. 135), if I have understood his point correctly, worries that Newton's third law of motion is violated during that process of collision. I am not sure that this is right, since the "rate of consumption" of dead force (along with the accumulation of living force, once it begins) is in balance, in the process described earlier. However, Suisky's worry highlights how difficult it is to make the details of Du Châtelet's complex account of forces explicitly consistent with Newton's laws of motion.

12 For discussion of 's Gravesande's epistemology and methodology, see Ducheyne (2014a and 2014b).

13 See Watkins (2005), pp. 27–8.

14 Du Châtelet (2018, 8.162) elaborates on what she means by mechanism and mechanical explanation using the clock metaphor, so loved by mechanical philosophers.

15 If the causal ordering of simples is such that bodies can "feel" the presence of one another at a distance, in virtue of their active and passive forces, then relational properties follow from the essence of body. A deeper exploration of action-at-a-distance in Du Châtelet's philosophy requires close attention to her account of essences, attributes, and modes (in Chapter 3), and I am grateful to Jamee Elder and Aaron Wells for our discussions of

this topic. Detlefsen (2014, p. 17) writes that Du Châtelet's commitment to mechanical explanation is consistent with PSR, but my point here is that it does not seem to follow from PSR (or from any other resources that she has available): it seems to be posited without supporting argument. This is all the more puzzling given Du Châtelet's engagement with the work of Maupertuis, who argued (1732, Chapter II) that the means by which bodies act on one another through impenetrability and contact action is no less mysterious to us than attraction.

16 Even Newtonian gravitational attraction falls into this category for Du Châtelet. Chapter 10 ends with a repeat of her commitment to a mechanical philosophy, as against Newtonian attraction (2018, 10.210):

> I will tell you in Chapter 16 how the Newtonians explain by attraction these same phenomena of cohesion, hardness, softness, and fluidity; for according to some among them, it is in these details that the necessity of admitting attraction is most manifest. Their Observations assuredly merit that we study them, and that we strive to find a mechanical reason for the Phenomena that they have observed.

17 The quotation continues with Du Châtelet's characteristic caution (2014, 9:182):

> Doubtless one must not abuse it, nor, in order to explain natural effects mechanically, invent motion and matter as one pleases… nor certainly without taking pains to demonstrate the existence of these matters and these motions. But neither must one limit nature to the number of fluids that we believe are needed for the explication of the phenomena, as several philosophers have done….

18 Du Châtelet takes Newtonian attraction and Locke's thinking matter to be examples of illegitimate appeals to the will of God (see 2009, 3.50). In principle, we eventually reach the end of explanation and must resort to God. But, for Du Châtelet, this is permissible only when we reach simple beings. Since a simple being can originate neither from another simple being nor from a compound, "it follows that the reason for these beings must be in the necessary being, that is to say, in God" (2009, 7.124, translation amended). This solves the puzzle set out in Musschenbroek (1744), where he asserts that we can never know whether or not we have reached the limit of our enquiries such that it is appropriate to appeal to the will of God.

19 In vortex theory, there is no empty space: there is matter everywhere. This is the plenum.

20 The following account is based on Brading (2018a).

21 As pointed out by Eric Schliesser in discussion, Du Châtelet's treatment of the empirical equivalence of the theories with respect to trajectories does not take into account comets, which will prove to be problematic for vortex theories. At the time, vortex accounts of comets had not yet been ruled out, but Maupertuis (in his "Letter upon comets" of 26 March 1742) wrote to Du Châtelet that comets prove the impossibility of vortices.

22 For Maupertuis's work on the shape of the Earth, see Beeson (1992). Du Châtelet's letters to Maupertuis are available online via ProjectVox.

23 Also relevant is her argument in Chapter 5 against the possibility of empty Newtonian absolute space and in favor of a plenum. This arguments uses

PSR, as well as the law of continuity (which in Chapter 1 she had argued follows from PSR). Insofar as Newtonian universal gravitation requires empty space, whereas vortex theory relies on a plenum, this argument supports vortex theory.

24 See Hutton (2004, pp. 521–2).

25 I draw on the analysis of Jamee Elder, who reconstructed valid arguments against Newtonian attraction from Du Châtelet's text and provided an assessment of their strengths and weaknesses. I am grateful to her and to Aaron Wells for our discussions of Du Châtelet on Newtonian gravitation.

26 See Hecht (2012, pp. 73–4) and Detlefsen (2014, Section 6, and forthcoming, p. 25 of the manuscript) for their discussions of Du Châtelet's arguments against Newtonian attraction. See Hutton (2004, p. 529) for Du Châtelet's modifications of her arguments concerning attraction in the 1742 edition of the *Foundations*.

27 It is Du Châtelet's work on *vis viva* that has received the most attention in the philosophical literature. See Iltis (1977, pp. 38–45), Hutton (2004, especially pp. 527–9), Hagengruber (2012b, pp. 35–8), Suisky (2012, pp. 144–6), Reichenberger (2012, pp. 157–71), Terrall (1995, p. 296–8), Kawashima (1990), and Walters (2001). See also Massimi and De Bianchi (2013) for the role of Du Châtelet's exchange with Mairan in Kant's early philosophy of matter and body.

28 Though it was a common trope to dismiss the *vis viva* controversy as a mere war of words, Reichenberger (2012) is correct in her assessment that it was "an examination of the ontological presuppositions underlying the metaphysics of the three dominant physico-philosophical theories of that time, the Cartesian, Leibnizian and Newtonian mechanics". For more on *vis viva* see Hankins (1965), Laudan (1968), Iltis (1970), Papineau (1977), Terrall (2004), Smith (2006), and references therein.

29 The argument from metaphysics given by Du Châtelet at this point concerns conservation of the total quantity of force in the universe.

30 The following discussion is drawn from a collaborative project I am working on with Marius Stan, concerning 18th-century matter theory and mechanics.

Appendix 1

Chapter Headings from Du Châtelet's *Foundations of Physics* and from Several Early 18th-Century Newtonian Textbooks

Du Châtelet (1740), *Foundations of Physics*, Volume I

Preface

1. Of the Principles of our Knowledge
2. Of the Existence of God
3. Of Essence, Attributes, and Modes
4. Of Hypotheses
5. Of Space
6. Of Time
7. Of the Elements of Matter
8. Of the Nature of Bodies
9. Of the Divisibility and Subtlety of Matter
10. Of the Shape and the Porosity of Bodies
11. Of Motion and Rest in General, and of Simple Motion
12. Of Complex Motion
13. Of Gravity
14. Of the Phenomena of Gravity, continued
15. Of Mr. Newton's Discoveries on Gravity
16. Of Newtonian Attraction
17. Of Rest, and the Fall of Bodies on an Inclined Plane
18. Of the Oscillation of Pendulums
19. Of the Movement of Projectiles
20. Of *Forces Mortes*, or Pressing Forces, and the Equilibrium of Forces
21. Of the Force of Bodies

Keill (1720), *An Introduction to Natural Philosophy* (from his 1700 Oxford lectures, first published in Latin in 1702)

Lecture I: Of the Method of Philosophizing
Lecture II: Of the Solidity and Extension of Bodies
Lecture III: Of the Divisibility of Magnitude
Lecture IV: Wherein the Objections Usually Brought against the Divisibility of Matter Are Answered
Lecture V: Of the Subtility of Matter
Lecture VI: Of Motion, Place and Time
Lecture VII: Definitions [of motion, speed, acceleration, momentum, impressed force, gravity, centripetal and centrifugal force, accelerating quantity of force]
Lecture VIII: [Philosophical Axioms]
Lecture IX: Theorems of the Quantity of Motion, and of Spaces Passed Over by Bodies in Motion
Lecture X. [continuation of the topic of Lecture IX]
Lecture XI. Of the Laws of Nature
Lecture XII. [continuation of the topic of Lecture XI]
Lecture XIII. Second Definitions [concerning centers of gravity]
Lecture XIV. [Elasticity]
Lecture XV. Of the Descent of Heavy Bodies on Inclined Planes, and of the Motion of Pendulums
Lecture XVI. [continuation of the topic of Lecture XV, including projectiles]

's Gravesande (1720), *Mathematical Elements of Physicks, Prov'd by Experiments: Being an Introduction to Sir Isaac Newton's Philosophy*, Book I (translated into English by John Keill from the 1720 Latin edition)

Chapter I. Concerning the Design of Physicks, and the Rules of Philosophizing
Chapter II. Of Body in General
Chapter III. Of Extension, Solidity, and Vacuity
Chapter IV. Of the Divisibility of Matter in Infinitum; and of the Subtlety of its Parts
Chapter V. Concerning the Cohesion of Parts, and also of Hardness, Softness, Fluidity, and Elasticity
Chapter VI. Of Motion in General, and also of Place and Time
Chapter VII. Of Comparing Motions One With Another
Chapter VIII. Of Comparing the Actions of Powers

Chapter IX. Generals Concerning Gravity
Chapter X. Of the Simple Trochlea, or Pulley; the Scale, or Balance; and of the Center of Gravity
Chapters XI–XV. [Of the Lever, Gears, Pulley, Wedge, Screw, and Compound Machines]
Chapter XVI. Concerning the Newtonian Laws of Nature
Chapter XVII. Of the Acceleration and Retardation of Heavy Bodies
Chapter XVIII. Of the Descent of Heavy Bodies on an Inclined Planes
Chapter XIX. Of the Oscillation of Pendulums
Chapter XX. Of Percussion, and the Communication of Motion
Chapter XXI. Of the Meeting and Striking Together of Elastick Bodies
Chapter XXII. Of Compound Motion, and Oblique Percussion
Chapter XXIII. Concerning Oblique Powers
Chapter XXIV. Of the Projection of Heavy Bodies
Chapter XXV. Of the Central Forces
Chapter XXVI. Of the Laws of Elasticity

(Book II concerns fluids, Book III fire and light, Book IV celestial gravitation)

Pemberton (1728), *A View of Sir Isaac Newton's Philosophy*

Introduction concerning Sir Isaac Newton's method of reasoning in philosophy

Book I

Chapter 1. Of the Laws of Motion
Chapter 2. Further Proofs of the Laws of Motion
Chapter 3. Of Centripetal Forces
Chapter 4. Of the Resistance of Fluids

Book II

Chapter 1. That the Planets Move in a Space Empty of Sensible Matter
Chapter 2. Concerning the Cause That Keeps in Motion the Primary Planets
Chapter 3. Of the Motion of the Moon and the Other Secondary Planets
Chapter 4. Of Comets
Chapter 5. Of the Bodies of the Sun and Planets
Chapter 6. Of the Fluid Parts of the Planets

Book III (Five chapters on optics)

Musschenbroek (1744), *The Elements of Natural Philosophy*, Volume 1 (translated from a later edition of the 1734 Latin publication)

Chapter I. Concerning Philosophy, and the Rules of Philosophizing
Chapter II. Of Body in General and Its Attributes
Chapter III. Of Vacuum, or Empty Space
Chapter IV. Of Place, Time, and Motion
Chapter V. Concerning Pressing Powers
Chapter VI. Of the Force of Bodies in Motion
Chapter VII. Concerning Gravity
Chapter VIII. Mechanics
Chapter IX. Of the Friction of Machines
Chapter X. Of Compound Motion
Chapter XI. Of the Descent of Heavy Bodies upon an Inclined Plane
Chapter XII. Of the Vibration of Pendulums
Chapter XIII. Of the Motion of Projectiles
Chapter XIV. Of Central Forces
Chapter XV. Of the Hardness, Softness, Fragility, Flexibility, and Elasticity of Bodies
Chapter XVI. Concerning Percussion
Chapter XVII. Of Electrical Bodies
Chapter XVIII. Of the Attractions of Bodies
Chapter XVIII. Of Coherence
Chapters XX–XXV. [These chapters all treat fluids]

References

*Works by Gabrielle Émilie le Tonnelier de Breteuil, marquise Du Châtelet, appear under the name "Du Châtelet, É."

Anstey, P. R. (2005). Experimental versus speculative natural philosophy. In P. R. Anstey & J. A. Schuster (Eds.), *The Science of Nature in the Seventeenth Century*, 215–42. Dordrecht: Springer.

Barber, W. H. (1967). Mme Du Châtelet and Leibnizianism: The genesis of the *Institutions de physique.* In William H. Barber, J. H. Brumfitt, R. A. Leigh, R. Shakleton, & S. S. B. Taylor (Eds.), *The Age of the Enlightenment: Studies Presented to Theodore Besterman*, 200–22. Edinburgh and London: University Court of the University of St. Andrews. Reprinted as Barber (2006).

Barber, W. H. (2006). Mme Du Châtelet and Leibnizianism: The genesis of the *Institutions de physique.* In J. P. Zinsser & J. C. Hayes (Eds.), *Emilie Du Châtelet: Rewriting Enlightenment Philosophy and Science.* Oxford: Voltaire Foundation.

Barbour, J. (2001). *The Discovery of Dynamics.* Oxford, New York: Oxford University Press.

Barfoot, M. (1990). Hume and the culture of science in the early eighteenth century. In M. A. Steward (Ed.), *Studies in the Philosophy of the Scottish Enlightenment*, 155–90. *Oxford Studies in the History of Philosophy.* Oxford: Clarendon Press.

Beeson, D. (1992). *Maupertuis: An Intellectual Biography. Studies on Voltaire and the Eighteenth Century* 299. Oxford: The Voltaire Foundation.

Bennett, J., & Remnant, P. (1978). How matter might at first be made. *Canadian Journal of Philosophy, 8*(suppl.), 1–11.

Berkeley, G. (1710). A treatise concerning the principles of human knowledge, Part I. In A. A. Luce & T. E. Jessop (Eds.), *The Works of George Berkeley, Bishop of Cloyne* (1948–1957) 2:41–113. London: Thomas and Nelson and Sons.

Berkeley, G. (1713). Three dialogues between Hylas and Philonous. In A. A. Luce & T. E. Jessop (Eds.), *The Works of George Berkeley, Bishop of Cloyne* (1948–1957) 2:163–263. London: Thomas and Nelson and Sons.

Berkeley, G. (1721). De motu. In A. A. Luce & T. E. Jessop (Eds.), *The Works of George Berkeley, Bishop of Cloyne* (1948–1957) 4:31–52. London: Thomas and Nelson and Sons.

Bernoulli, J. I. (1735). *Essai d'une nouvelle Physique céleste*: Imprim. Royale.

Biener, Z. (2018). Newton's Regulae Philosophandi. In E. Schliesser & C. Smeenk (Eds.), *The Oxford Handbook of Newton.* doi:10.1093/oxfordhb/9780199930418.013.4.

Biener, Z., & Smeenk, C. (2012). Cotes' queries: Newton's empiricism and conceptions of matter. In A. Janiak & E. Schliesser (Eds.), *Interpreting Newton: Critical Essays.* 105–37. Cambridge: Cambridge University Press.

Brading, K. (2018a). Émilie Du Châtelet and the problem of bodies. In E. Thomas (Ed.), *Early Modern Women on Metaphysics*, 150–68. Cambridge University Press.

Brading, K. (2018b). Newton on body. In E. Schliesser & C. Smeenk (Eds.), *The Oxford Handbook of Newton.* doi:10.1093/oxfordhb/9780199930418.013.10.

Brading, K. (forthcoming). A note on rods and clocks in Newton's *Principia. Studies in History and Philosophy of Modern Physics.* doi:10.1016/j.shpsb.2017.07.004.

Carboncini, S. (1987). L'Encyclopédie et Christian Wolff: a propos de quelques articles anonymes. *Les études philosophiques, 4*, 489–504.

Cohen, I. B. (1978). *Introduction to Newton's 'Principia'.* Cambridge, MA: Harvard University Press.

Descartes, R. (1991). *Principles of Philosophy*, trans. V. R. Miller & R. P. Miller. Dordrecht, Boston, London: Kluwer Academic Publishers.

Detlefsen, K. (2014). Émilie du Châtelet. In E. N. Zalta (Ed.), *The Stanford Encyclopedia of Philosophy* (Summer 2014 Edition). Retrieved from https://plato.stanford.edu/archives/sum2014/entries/emilie-du-chatelet/.

Detlefsen, K. (forthcoming). Du Châtelet and Descartes on the role of hypothesis and metaphysics in science. In E. O'Neill & M. Lascano (Eds.), *Feminist History of Philosophy: The Recovery and Evaluation of Women's Philosophical Thought.* Springer.

DiSalle, R. (2006). *Understanding Space-Time: The Philosophical Development of Physics from Newton to Einstein.* Cambridge: Cambridge University Press.

Du Châtelet, É. (1738). Lettre sur les Eléments de la Philosophie de Newton. *Journal des sçavans*, 458–75.

Du Châtelet, É. (1739). Dissertation sur la nature et la propagation du feu. In *Recueil des pièces qui ont remporté le prix de l'Académie royale des Sciences en 1738*, ed. Académie royale des Sciences, 85–168. Paris: Imperimerie royale. For extracts in English translation see Zinsser and Bour (2009).

Du Châtelet, É. (1740). *Institutions de physique.* Paris: Prault.

Du Châtelet, É. (1742). *Institutions physiques de Madame la marquise Du Chastelet, addressés à Mr. son fils.* Amsterdam: Aus dépens de la Compagnie.

Du Châtelet, É. (2009). *Foundations of Physics*, extracts. In Zinsser & Bour (2009), 115–200.

Du Châtelet, É. (2014). *Foundations of Physics*, Chapter 9. In Patton (2014), 332–42.

Du Châtelet, É. (2018). *Foundations of Physics*, extracts. Trans. K. Brading et al. Retrieved from www.kbrading.org/du-chatelet.

Ducheyne, S. (2014a). 's Gravesande's appropriation of Newton's natural philosophy, Part I: epistemological and theological issues. *Centaurus, 56*(1), 31–55.

Ducheyne, S. (2014b). 's Gravesande's appropriation of Newton's natural philosophy, Part II: methodological issues. *Centaurus, 56*(2), 97–120.

Ducheyne, S. (2015). Petrus van Musschenbroek and Newton's 'vera stabilisque Philosophandi methodus'. *Berichte zur Wissenschaftsgeschichte, 38*, 279–304.

Ehrman, E. (1986). *Madame du Châtelet: Scientist, Philosopher and Feminist of the Enlightenment*. Leamington Spa: Berg Publishers.

Euler, L. (1736). *Mechanica sive motus scienta analytice exposita... instar supplementi ad Commentar. Acad. scient. imper*: ex typographia Academiae scientarum.

Fara, P. (2004). *Pandora's Breeches: Women, Science and Power in the Enlightenment*. London: Pimlico.

Gale, G. (1970). The physical theory of Leibniz. *Studia Leibnitiana* (H. 2), 114–27.

Garber, D. (1992). *Descartes' Metaphysical Physics*. Chicago and London: University of Chicago Press.

Garber, D. (2009). *Leibniz: Body, Substance, Monad*. Oxford and New York: Oxford University Press.

Gueroult, M. (1980). The metaphysics and physics of force in Descartes. In S. Gaukroger (Ed.), *Descartes: Philosophy, Mathematics and Physics*, 196–229. Brighton: Harvester Press.

Hagengruber, R. (Ed.) (2012a). *Emilie du Châtelet between Leibniz and Newton*. Dordrecht, New York: Springer.

Hagengruber, R. (2012b). Emilie du Châtelet between Leibniz and Newton: The Transformation of Metaphysics. In Hagengruber (2012a), 1–59.

Hankins, T. L. (1965) Eighteenth-Century Attempts to Resolve the Vis Viva Controversy. *Isis 56*(3), 281–97.

Harper, W. L. (2011). *Isaac Newton's Scientific Method*. Oxford: Oxford University Press.

Harth, E. (1992). *Cartesian Women: Versions and Subversions of Rational Discourse in the Old Regime*. Italica, London: Cornell University Press.

Hecht, H. (2012). *In the Spirit of Leibniz–Two Approaches from 1742*. In Hagengruber (2012a), 61–75.

Henry, J. (2011). Galileo and the scientific revolution: The importance of his kinematics. *Galileana, XVIII*, 3–36.

Holden, T. (2004). *The Architecture of Matter: Galileo to Kant*. Oxford: Oxford University Press.

Hume, D. (1748). *Philosophical Essays Concerning Human Understanding* (later reprinted as *An Enquiry Concerning Human Understanding*). London: A. Millar.

Hume, D. (1939–40). *A Treatise of Human Nature*. London: John Noon.

Hutton, S. (2004). Emilie du Châtelet's Institutions de physique as a Document in the History of French Newtonianism. *Studies in History and Philosophy of Science Part A, 35*(3), 515–31.

Hutton, S. (2012). Between Newton and Leibniz: Emilie du Châtelet and Samuel Clarke. In Hagengruber (2012a), 77–95. Springer.

Huygens, C. (1690). *Traité de la Lumière*. Leyden: Pieter van der Aa.

Huygens, C. (2009). Successful hypotheses and high probability (extract from his *Treatise on Light*). In T. McGrew, M. Alpsector-Kelly, & F. Allhoff (Eds.), *Philosophy of Science: An Historical Anthology*, pp. 164–6. Oxford: Wiley-Blackwell.

Iltis, C. (1970). D'Alembert and the vis viva controversy. *Studies in History and Philosophy of Science A, 1*(2), 135–44.

Iltis, C. (1977). Madame du Châtelet's metaphysics and mechanics. *Studies in History and Philosophy of Science Part A, 8*(1), 29–48.

Janik, L. G. (1982). Searching for metaphysics of science: The structure and composition of Madame du Chatelet's Institutions de physique, 1737–1740. *Studies on Voltaire, 201*, 85–113.

Kant, I. (1992). *Theoretical Philosophy 1755–1770*. D. Walford in collaboration with R. Meerbote (Eds. and trans.), *The Cambridge Edition of the Works of Immanual Kant*. Cambridge, New York: Cambridge University Press.

Kant, I. (2012a). *Kant: Natural Science*. E. Watkins (Ed.), L. W. Beck, J. B. Edwards, O. Reinhardt, M. Schönfeld, & E. Watkins (Trans.), *The Cambridge Edition of the Works of Immanual Kant*. Cambridge, New York: Cambridge University Press.

Kant, I. (2012b). Thoughts on the true estimation of living forces. In Kant (2012a), 1–155.

Kawashima, K. (1990). La participation de madame du Châtelet à la querelle sur les forces vives. *Historia Scientiarum: International Journal of the History of Science Society of Japan, 40*, 9–28.

Kawashima, K. (2004). Birth of ambition: Madame du Chatelet's Institutions de physique. *Historia Scientiarum. Second Series: International Journal of the History of Science Society of Japan, 14*(1), 49–66.

Keill, J. (1702). *Introductio ad veram physicam Seu lectiones physicæ. Habitæ in schola naturalis philosophiæ Academiæ Oxoniensis, quibus accedunt Christiani Hugenii theoremata de vi centrifuga & motu circulari demonstrata, per Jo. Keill*. London: Thomas Bennett.

Keill, J. (1720). *An Introduction to Natural Philosophy: Or, Philosophical Lectures Read in the University of Oxford, Anno Dom. 1700. To Which Are Added, the Demonstrations of Monsieur Huygens's Theorems, Concerning the Centrifugal Force and Circular Motion*. London: H.W.

Lascano, M. P. (2011). Emilie du Châtelet on the existence and nature of god: An examination of her arguments in light of their sources. *British Journal for the History of Philosophy, 19*(4), 741–58.

Laudan, L. L. (1968). The vis viva controversy, a post-mortem. *Isis, 59*(2), 130–43.

Lee, S. (2016). Occasionalism. In E. N. Zalta (Ed.), *The Stanford Encyclopedia of Philosophy* (Winter 2016 Edition). Retrieved from: https://plato.stanford.edu/archives/win2016/entries/occasionalism/.

Leibniz, G. W. F., & Clarke, S. (1998). *The Leibniz-Clarke Correspondence: Together with Extracts from Newton's Principia and Opticks*. H. G. Alexander (Ed.), Manchester: Manchester University Press.

Locke, J. (1689). *An Essay Concerning Human Understanding*. Thomas Bassett: London.

Maglo, K. (2008). Mme Du Châtelet, l'Encyclopédie, et la philosophie des sciences. In U. Kölving & O. Courcelle (Eds.), *Émilie Du Châtelet: Éclairages & Documents Nouveaux*, 255–66. Paris, Ferney-Voltaire: CIEDS.

Massimi, M., & De Bianchi, S. (2013). Cartesian echoes in Kant's philosophy of nature. *Studies in History and Philosophy of Science Part A, 44*(3), 481–92.

Maudlin, T. (2012). *Philosophy of Physics: Space and Time*. Princeton, NJ: Princeton University Press.

Maupertuis, P.-L. M. (1732). Sur les loix d'attraction. *Memoires of the Royal Academy of Science*, 343–63.

Maupertuis, P.-L. M. (1742). *Lettre sur la comète qui paroissoit en 1742*. Paris.

Musschenbroek, P. van (1734). *Elementa physicae*. Leyden: Samuel Luchtmans.

Musschenbroek, P. van (1739). *Essai de physique*. Leyden: Samuel Luchtmans.

Musschenbroek, P. van (1744). *The Elements of Natural Philosophy*, trans. J. Colson. J. Nourse: London.

Musschenbroek, P. van (1748). *Institutiones physicae conscriptae in usus academicos*. Lugduni Batavorum: Samuel Luchtmans et Filium.

Newton, I. (1999). *The Principia: Mathematical Principles of Natural Philosophy*, trans. I. B. Cohen and A. Whitman. Berkeley, Los Angeles, London: University of California Press.

O'Neill, E. (1998). Disappearing ink: Early modern women philosophers and their fate in history. In J. Kourany (Ed.), *Philosophy in a Feminist Voice: Critiques and Reconstructions*, 17–62. Princeton, NJ: Princeton University Press.

Papineau, D. (1977). The *vis viva* controversy: do meanings matter?. *Studies in History and Philosophy of Science A, 8*, 111–142.

Patton, L. (2014). *Philosophy, Science, and History: A Guide and Reader*. New York, London: Routledge.

Pemberton, H. (1728). *A View of Isaac Newton's Philosophy*. Dublin: John Hyde, and John Smith and William Bruce.

Reichenberger, A. (2012). Leibniz's quantity of force: A 'Heresy'? Émilie du Châtelet's institutions in the context of the Vis Viva controversy. In Hagengruber (2012a), 157–71.

Reichenberger, A. (2018). Émilie Du Châtelet's interpretation of the laws of motion in the light of 18th century mechanics. *Studies in History and Philosophy of Science A, 69*, 1–11.

Rey, A.-L. (forthcoming). *Les Certitudes des Lumières*. Publisher: Classiques Garnier.

Rodrigues, A. (2012). Emilie Du Châtelet, a Bibliography. In Hagengruber (Ed.), 207–46.

Roe, G. (2018). A Sheep in Wolff's Clothing: Émilie du Châtelet and the Encyclopédie. *Eighteenth-Century Studies, 51*(2), 179–96.

Rohault, J. (1671). *Traité de Physique*. Paris: Savreux.

Rutherford, D. (2004). Idealism declined: Leibniz and Christian Wolff. In P. Lodge (Ed.), *Leibniz and his Correspondents*, 214–37. Cambridge: Cambridge University Press.

Schönfeld, M. (2000). *The Philosophy of the Young Kant: The Precritical Project.* Oxford: Oxford University Press.

's Gravesande, W. J. (1720). *Mathematical Elements of Physicks, Prov'd by Experiments: Being an Introduction to Sir Isaac Newton's Philosophy*, Book I (translated into English by John Keill from the 1720 Latin edition).

Smith, G. E. (2002). The methodology of the *Principia*. In I. B. Cohen and G. E. Smith (Eds.), *The Cambridge Companion to Newton*, 138–73. Cambridge: Cambridge University Press.

Smith, G. E. (2006). The vis viva dispute: A controversy at the dawn of dynamics. *Physics Today, 59*(10), 31.

Stan, M. (2018). Emilie du Châtelet's metaphysics of substance. *Journal of the History of Philosophy, 56*(3), 477–96.

Suisky, D. (2012). Leonhard Euler and Emilie du Châtelet. On the Post-Newtonian Development of Mechanics. In R. Hagengruber (2012a), 113–55.

Terrall, M. (1995). Emilie du Châtelet and the gendering of science. *History of Science, 33*(3), 283–310.

Terrall, M. (2002). *The Man Who Flattened the Earth: Maupertuis and the Sciences in the Enlightenment.* Chicago: University of Chicago Press.

Terrall, M. (2004). Vis viva revisited. *History of Science, 42*(2), 189–209.

Van Strien, M. (2014). On the origins and foundations of Laplacian determinism. *Studies in History and Philosophy of Science Part A, 45*, 24–31.

Vanzo, A. (2015). Christian Wolff and experimental philosophy. *Oxford Studies in Early Modern Philosophy, 7*, 225–55.

Voltaire (1738). *The Elements of Sir Isaac Newton's Philosophy.* Stephen Austin: London.

Voltaire (1989). *The Complete Works of Voltaire* 14. In W. H. Barber & U. Kolving (Eds.), Oxford: Voltaire Foundation.

Wade, I. O. (1969). *The Intellectual Origins of Voltaire.* Princeton, NJ: Princeton University Press.

Walters, R. L. (2001). La querelle des forces vives et le rôle de Mme du Châtelet. In F. de Gandt (Ed.) *Cirey dans la vie intellectuelle: la réception de Newton en France*, 198–211. Oxford: Voltaire Foundation

Watkins, E. (2005). *Kant and the Metaphysics of Causality.* Cambridge: Cambridge University Press.

Watkins, E. (2006). On the Necessity and Nature of Simples: Leibniz, Wolff, Baumgarten, and the Pre-Critical Kant. In D. Garber & S. Nadler (Eds.), *Oxford Studies in Early Modern Philosophy Volume 3*, 261–341. Oxford: Clarendon Press.

Winter, U. (2012). From translation to philosophical discourse–Emilie Du Châtelet's commentaries on Newton and Leibniz. In Hagengruber (2012a), 173–206.

Wolff, C. (1729). De Hypothesibus philosophicis. In his *Horae subsecivae marburgenses anni MDCCXXIX*, 177–230. Frankfurt a.M.: Renger.

Wolff, C. (1963). *Preliminary Discourse on Philosophy in General.* R. J. Blackwell (trans.). Indianapolis, New York: The Bobbs-Merrill Company, Inc.

Wolff, C. (2009). Rational thoughts on god, the world and the soul of human beings, also all things in general. In E. Watkins (Ed. and trans.), *Kant's Critique of Pure reason: Background Source Materials* (2009), 7–53. Cambridge University Press.

Zinsser, J. P. (2006). *Emilie Du Châtelet: Daring Genius of the Enlightenment.* Penguin. First published as *La dame d'esprit: a biography of the Marquise Du Châtelet* by Viking Penguin.

Zinsser, J. P., & Bour, I. (2009). *Emilie Du Châtelet. Selected Philosophical and Scientific Writings.* Chicago: University of Chicago Press.

Zinsser, J. P., & Hayes, J. C. (2006). (Eds.) *Emilie Du Châtelet: Rewriting Enlightenment Philosophy and Science.* Oxford: Voltaire Foundation.

Index

CPSIA information can be obtained
at www.ICGtesting.com
Printed in the USA
BVHW012203011022
648485BV00003B/75

9 781138 351653